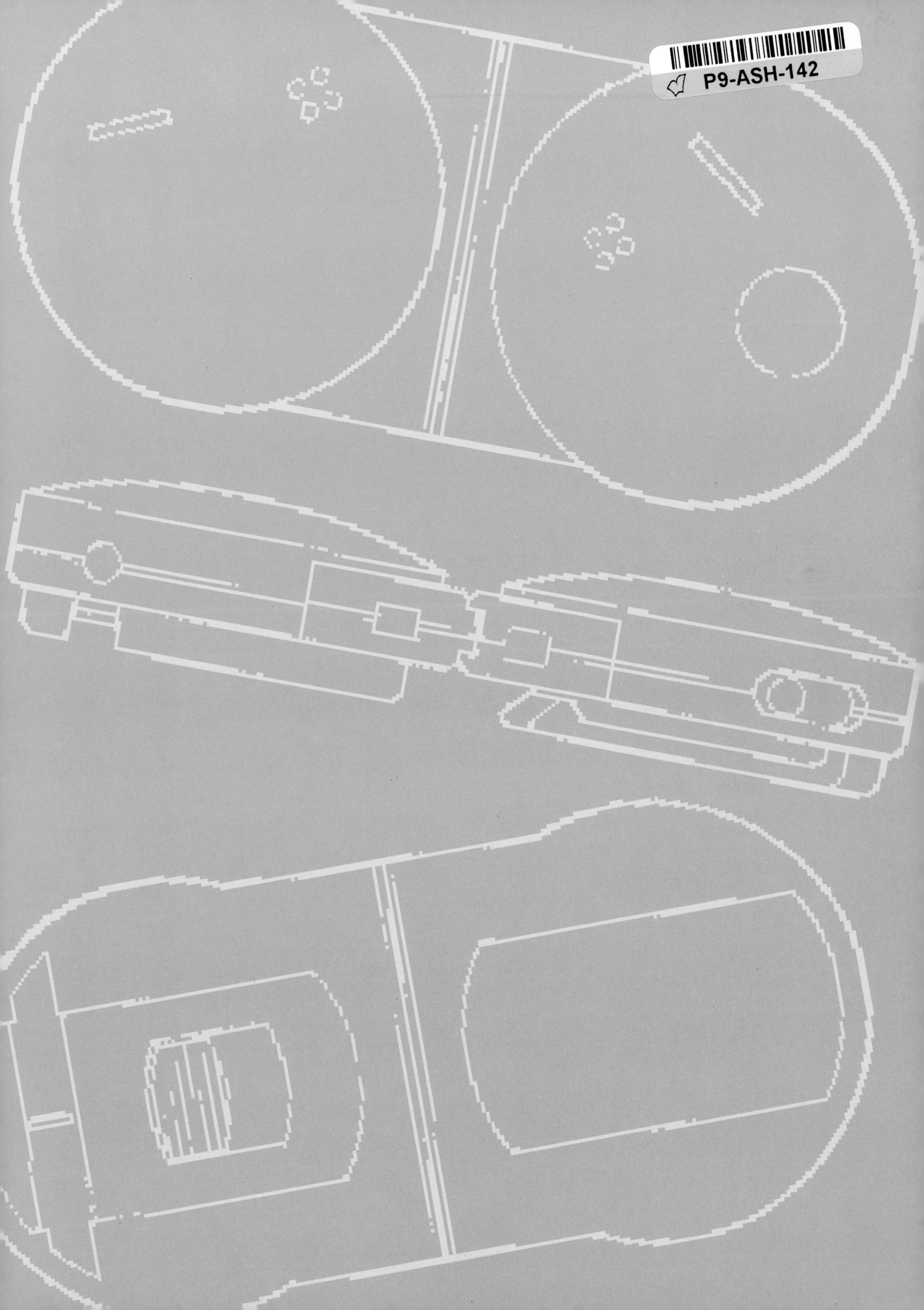

Innovation

Award-Winning Industrial Design

IDSA
Industrial Designers Society of America

Library of Applied Design

An Imprint of
PBC International, Inc.

Distributor to the book trade in the United States and Canada
Rizzoli International Publications Inc.
300 Park Avenue South
New York, NY 10010

Distributor to the art trade in the United States and Canada
PBC International, Inc.
One School Street
Glen Cove, NY 11542

Distributor throughout the rest of the world
Hearst Books International
1350 Avenue of the Americas
New York, NY 10019

Library of Congress Cataloging-in-Publication Data

Innovation: Award-winning industrial design / by the Industrial Designers Society of America
p. cm.
Includes index.
ISBN 0-86636-332-7 (Pbk ISBN 0-86636-377-7)
1. Design, Industrial-United States-Awards. I. Industrial Designers Society of America.
TS23.I66 1994
745.2'079'73-dc20
94-9007
CIP

Caveat- Information in this text is believed accurate, and will pose no problem for the student or casual reader. However, the author was often constrained by information contained in signed release forms, information that could have been in error or not included at all. Any misinformation (or lack of information) is the result of failure in these attestations. The author has done whatever is possible to insure accuracy.

Color separation by
Fine Arts Repro House Co., Ltd., H.K.

Printing and binding by
Toppan Printing Co., (H.K.) Ltd.

Printed in China

10 9 8 7 6 5 4 3 2 1

Dedicated to industrial designers:

Their innovation makes the world a little better each day.

table of Contents

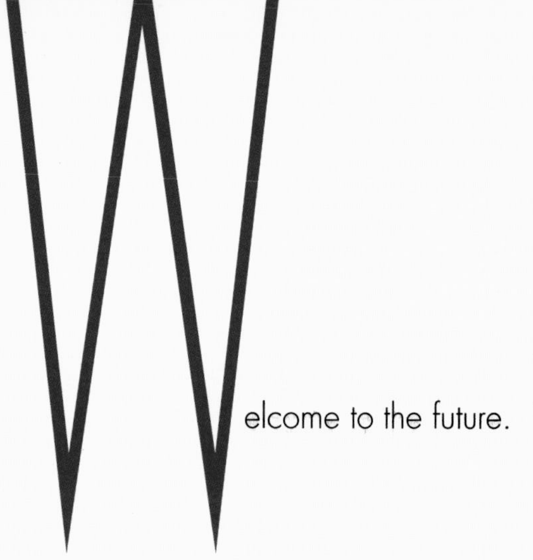

Welcome to the future. The designs shown on the following pages provide a glimpse of what lies ahead for designers and for the rest of us.

Give yourself the freedom to wander through these pages. Take the time to settle on images and concepts that you find especially interesting. Ideas and images are presented in a loose format that invites the same reflection and quiet appreciation as a gallery or library. You will learn something new every time you return.

"In modern *design, information is more important than art."*

I encourage you to look beyond the visual designs to the thought behind each idea and innovation. Design is changing. Art and information are converging to create products that are useful and human.

Design is expanding. Visual detailing is becoming a smaller and smaller part of what design is all about. Designers are going beyond product appearance to become more involved in function and process. Designers are working with the *soul* of the product.

"Good design *adds value without adding cost."*

This change in focus is documented here. From tire treads to automobile engines, designers are starting to shape products in new and unique ways. Our work product is gradually shifting from drawings and images to concepts and ideas. We expect this trend to continue.

"Fit, form, *function and fun.*"

Yesterday, designers were asked to create image. Today, we are expected to create value.

The accelerating pace of technological change is creating a related need for humanity. The concept of *product personality* is taking on new meaning and value in a world that seems a little bit less personal everyday.

The human side of design should not be forgotten. From ease of use to *products for everyone*, designers are renewing their commitment to the people at the end of the process: customers. Products are more than *things*. Products are human experiences.

"Informed *creativity.*"

The work shown on the following pages is the result of art, effort and information. Enjoy your journey through these pages.

Sohrab Vossoughi
President, ZIBA Design, Inc.

Foreword

If there is one idea that captures the American spirit, it is innovation. I mean that term in its broadest sense, in the sense of finding a solution in order to better the situation. It finds its expression not only in technology and products, but in how we think of ourselves and how we act as a society.

Unlike all other civilizations, we are convinced that any obstacle can be overcome, any ill cured, any goal achieved. It's not the work ethic that is uniquely American. It is this idea that, if only we apply enough ingenuity, we can walk on the moon, cure cancer, end war and feed everyone. If we haven't cured cancer yet, we have fueled a technological revolution that has spurred economic activity for more than a century.

We are relentless in our pursuit and celebration of innovation. We lionize it in movies, books and articles. We pride ourselves on leading the world in research. (Not in the amount of money we invest, but in the breakthroughs we achieve.) We are even beginning to nurture it from the cradle up through educational programs that help children apply their innate creativity.

Clearly, we are not the only country that is innovative. Too many breakthroughs in too many other countries prove that. But only in the US has innovation—or invention—been raised to the level of Muse.

In the heart of every industrial designer lies a longing to innovate, to design a better mousetrap and catch the brass ring of success. Entrepreneurship based on innovation is an idea that makes every industrial designer's pulse race.

It is no coincidence, then, that innovation has always topped the list of criteria by which jurors evaluate entries in the Industrial Design Excellence Awards program. Whether the innovation is as grand as a completely new product or as small as a better handle, innovation is the special quality that, even more than aesthetics, the jury looks for in winners. And that has held true in every jury since 1981.

Although innovation is not unique to industrial designers, industrial designers have a special brand of innovation born of a facility for making connections between disparate things. Consider the idea of applying the inflatable bladder technology from the medical industry to the problem of making a sneaker—and later a baseball mitt—fit better.

Design innovation may entail breaking with visual stereotypes or eliminating screws or using a new material. It almost always helps to improve functionality by making a product comfortable, easy and pleasant to use, safe and durable. In the past few years, the need to design products that are ecologically responsible that use less material or recycled material, that are easy to service and then disassemble for recycling—has emerged to fuel design innovation.

The climate for industrial design innovation has improved over the past few years as US business has become more aware of this ingredient in achieving customer satisfaction. And that improvement will escalate in the next few years, as the US moves to a predominantly civilian economy and away from the heavy investments in military spending of the past 50 years. Opportunities for including industrial design in the effort to develop commercially viable products will accelerate through such efforts as TAP-IN, a federally funded program in which IDSA will connect industrial designers with manufacturers, federal laboratories and regional technology transfer centers. This view of industrial designers as contributors who can conceive of new and different applications for technology will help US industry transition and regain the competitive edge.

Other developments, currently in their infancy, promise to burgeon into policy-level recognition and institutionalization of industrial design. Congress is looking at a bill that would establish a US Design Council in the US Department of Commerce, and the National Endowment for the Arts has developed a proposal for a White House Council on Design. If implemented, the two bodies will put design where it has never been before, institutionalizing it in positions of power and influence. Regardless of the ultimate outcome, the very fact that these measures are under discussion at the highest levels of national policy will produce valuable connections and opportunities for participation that never existed before. The result can only be more well-designed products and greater product innovation.

And so it is timely that there be a book devoted to the celebration of design innovation. Given the preeminent position held by innovation in the criteria of the Industrial Design Excellence Awards (IDEA) program, that competition provides a wonderful source of material for such a book.

Introduction

The designs in this book have won either a Gold, Silver or Bronze award in the 1991, 1992 and 1993 IDEA competition. Sponsored by the Industrial Designers Society of America and *Business Week* magazine, the IDEA program was developed in 1980 with funding from the National Endowment of the Arts.

One of the tenets of the competition has been for jurors to objectively evaluate each entry using carefully crafted, published criteria of excellence. Those criteria have evolved in the past few years to encompass ecological responsibility. The five criteria are now: design innovation; benefit to user; benefit to client/manufacturer; ecological responsibility of materials and processes throughout life cycle; and visual appeal.

Every entry is rated by a juror on how well it achieves these criteria. Those entries with the highest scores become the finalists that the juror presents to the jury as a whole, recommending each for a level of award and explaining the rationale. The jury discusses these recommendations and votes on the level of award, if any, to give to each finalist.

Over the years, the Industrial Design Excellence Awards program has earned an unsurpassed reputation for objectivity and credibility. The tools used by the jury are an important part of its credibility. Although for logistical reasons the actual products are not submitted, the entrants are asked to supply photos and videos of the products in use, along with schematics, exploded views and summaries of related research. In addition, the entrants provide essay answers to questions that ask for descriptions of the design problem and its solution, relating these to the criteria.

The jury itself is made up of nationally prominent US industrial designers who waive the right to enter the competition in the year they serve on the jury. This dedication to helping their profession find and celebrate excellence reflects the commitment, activism and volunteer spirit of IDSA's 2,250 members.

IDSA is a nonprofit organization dedicated to improving industrial design knowledge and to representing the profession to business, government and the public. Established in 1965, IDSA today serves the profession by publishing a directory, journal and newsletter; holding national and district conferences and chapter activities; reaching out to the press and government and educational institutions; and conducting the IDEA. Its committees address such far-reaching topics as diversity, universal design and ecological responsibility, and its sections provide information in such diverse areas as women in design, human factors, visual interface, housewares and furniture.

With innovation so central to all the winners, it is only appropriate that this book itself be somewhat innovative. In an unusual departure for such a design book, each of the designs have descriptive text. The Gold winners, in particular, are presented with design problem, solution and results. The book is organized with Section I devoted to Designing for the Consumer; Section II devoted to Designing for Industry; and Section III devoted to Environmental and Packaging Design.

As with any broad area of human endeavor, innovation is not always wonderful. Sometimes an improvement in one area brings with it an unexpected problem in another. Big, oval handles are easier to grip but use more materials, pitting benefits to the user against ecological responsibility.

This book celebrates innovation but cannot address all the implications of its many occurrences. Sometimes you simply have to sit back and let the wonder of human ingenuity and its boundless optimism woo you. Enjoy!

KRISTINA GOODRICH
SENIOR DIRECTOR OF EXTERNAL AFFAIRS, IDSA

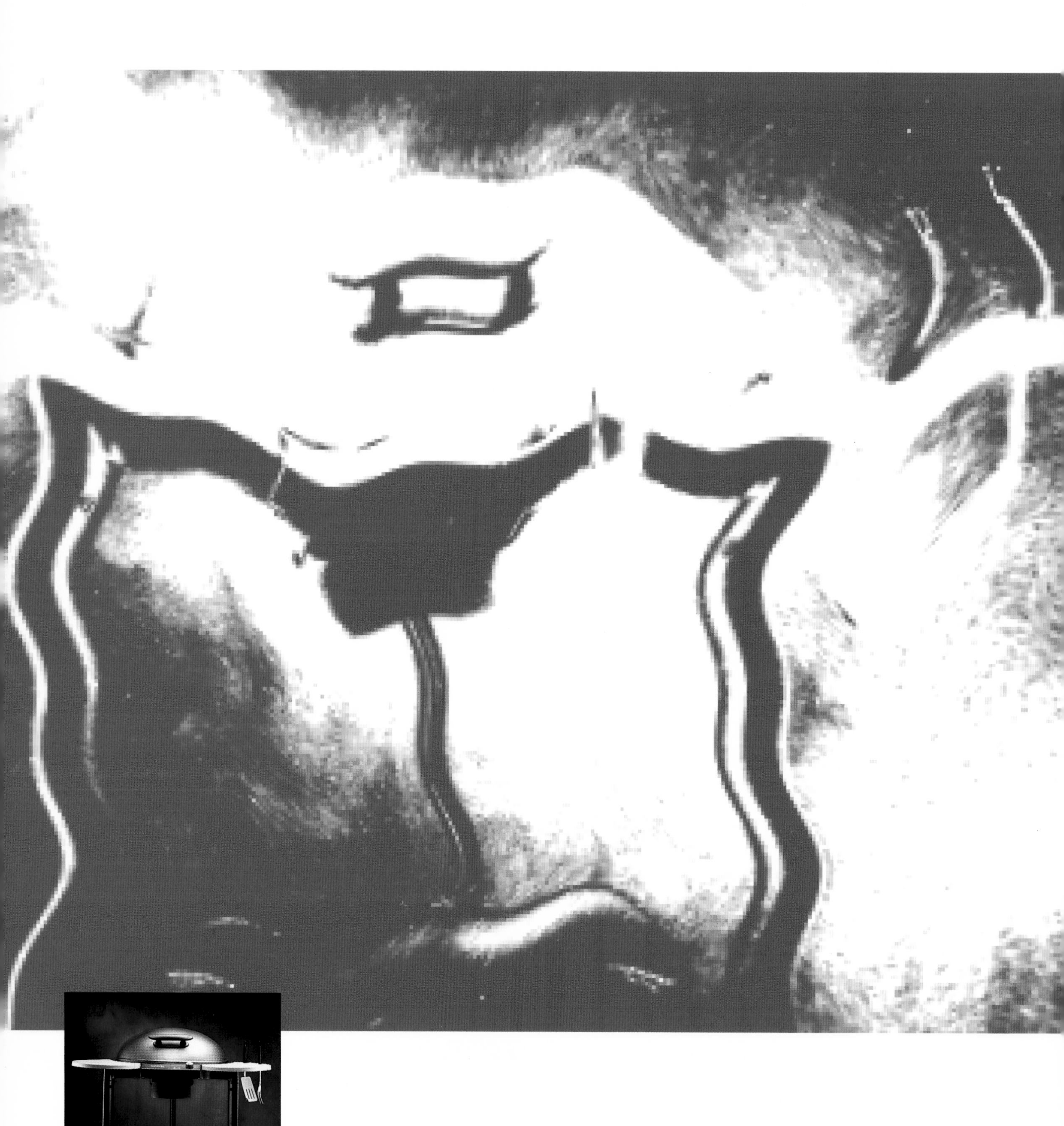

designing for the

Consumer

EZ Router™

Design Objective

To improve router performance, simplify depth adjustment, enhance product features, and extend the router's service life.

Design Solutions

The MicroDepth adjustment ring concept allows full depth adjustment simply with one revolution of the ring. Designed to interlock with the motor housing and the base, the MicroDepth ring prevents the motor housing from slipping during depth adjustment, especially when used with a router table. Comfortable handles and a convenient switch location further improve this router over its predecessor.

Result

The EZ Router's design radically improves such user concerns as ease of use and safety. Compared to other routers, it is dramatically less cumbersome.

Other Awards

–Sears Product Innovation Award

"The router's central, visual theme–the red ramp–immediately tells the user where and how to adjust the cutter depth.... This is design–not just styling."

JUROR STEPHEN HAUSER, IDSA

Designers

Bob McCracken, Dave Beth of Ryobi Motor Products Corp.;
Jim Watson, Karen Wilk, Alex Chunn of Industrial Design Associates

Client

Sears Roebuck & Co.

BeeperKid

DESIGNERS
Lev Chapelsky; Kennedy Design

CLIENT
A + H International, Inc.

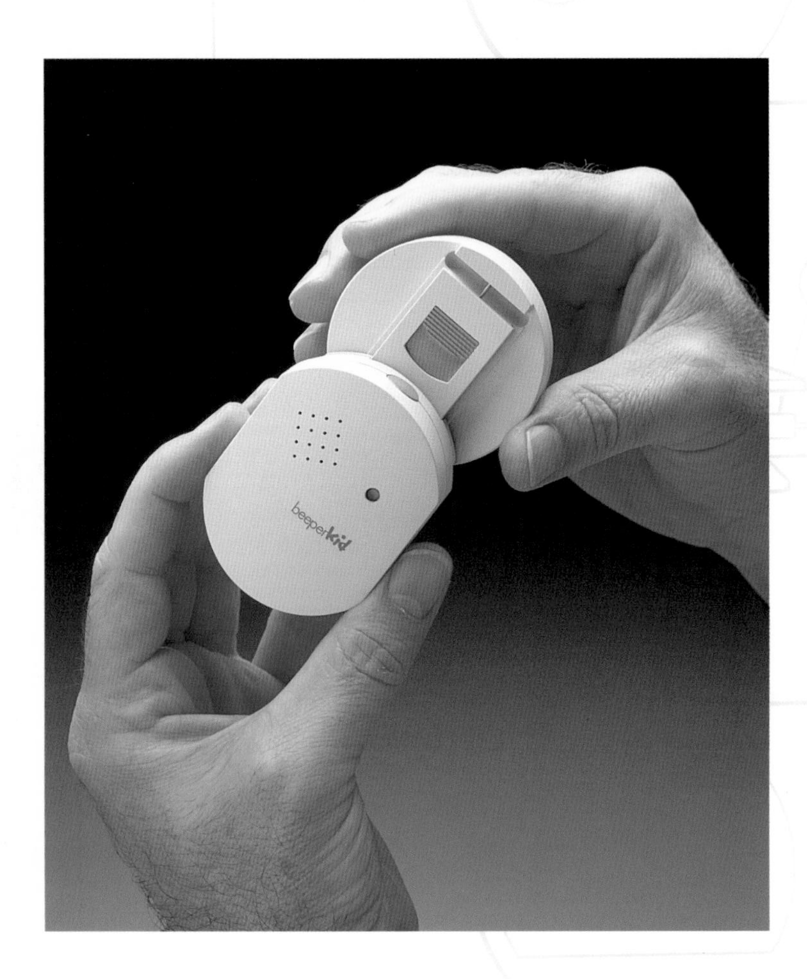

"The BeeperKid demonstrates the powerful benefits that the industrial design process can offer. The client provided the technology and relied on the industrial designer to provide the idea for the product's form and how it should be used."

JUROR LIZ POWELL

"...The design is metaphorically and functionally ingenious: at home both parent and child disks live together as one on a recharger pedestal and click on automatically as they are pulled apart."

TIME, JANUARY 4, 1993

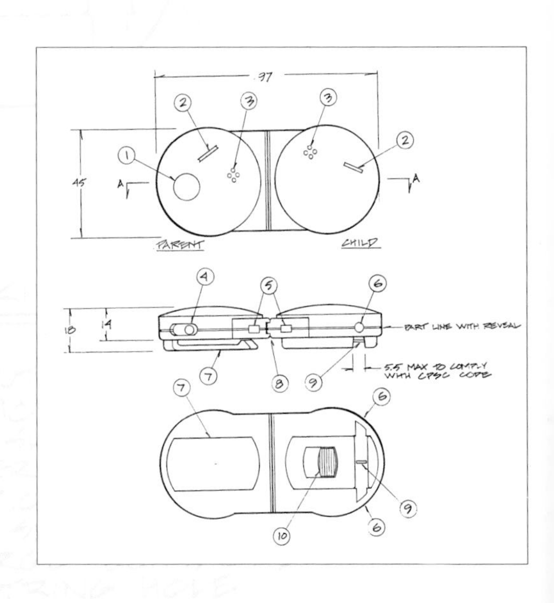

DESIGN OBJECTIVE

To design a product that signals the parent when a child wanders too far away.

DESIGN SOLUTIONS

When the parent and child units are removed from the recharging pedestal and separated, the system automatically turns on and the monitoring begins. This integration of the on/off switches with the interlock makes the units foolproof for the parent and childproof to the kid. The recharging pedestal, which eliminated the need for batteries, adds convenience, reliability, and safety.

RESULT

When a child goes beyond a certain distance, an alarm sounds on the parent's device; then the parent can sound a signal on the child's device, a beacon for locating the child. Originally manufactured for market release in 1992, the revised product, with a unique technology licensed by the military and previously unavailable to commercial business, was scheduled for release in August 1994.

OTHER AWARDS

–*Time*, The Best of Design of 1992
–Electronic Industries Association, Innovations '92 Award

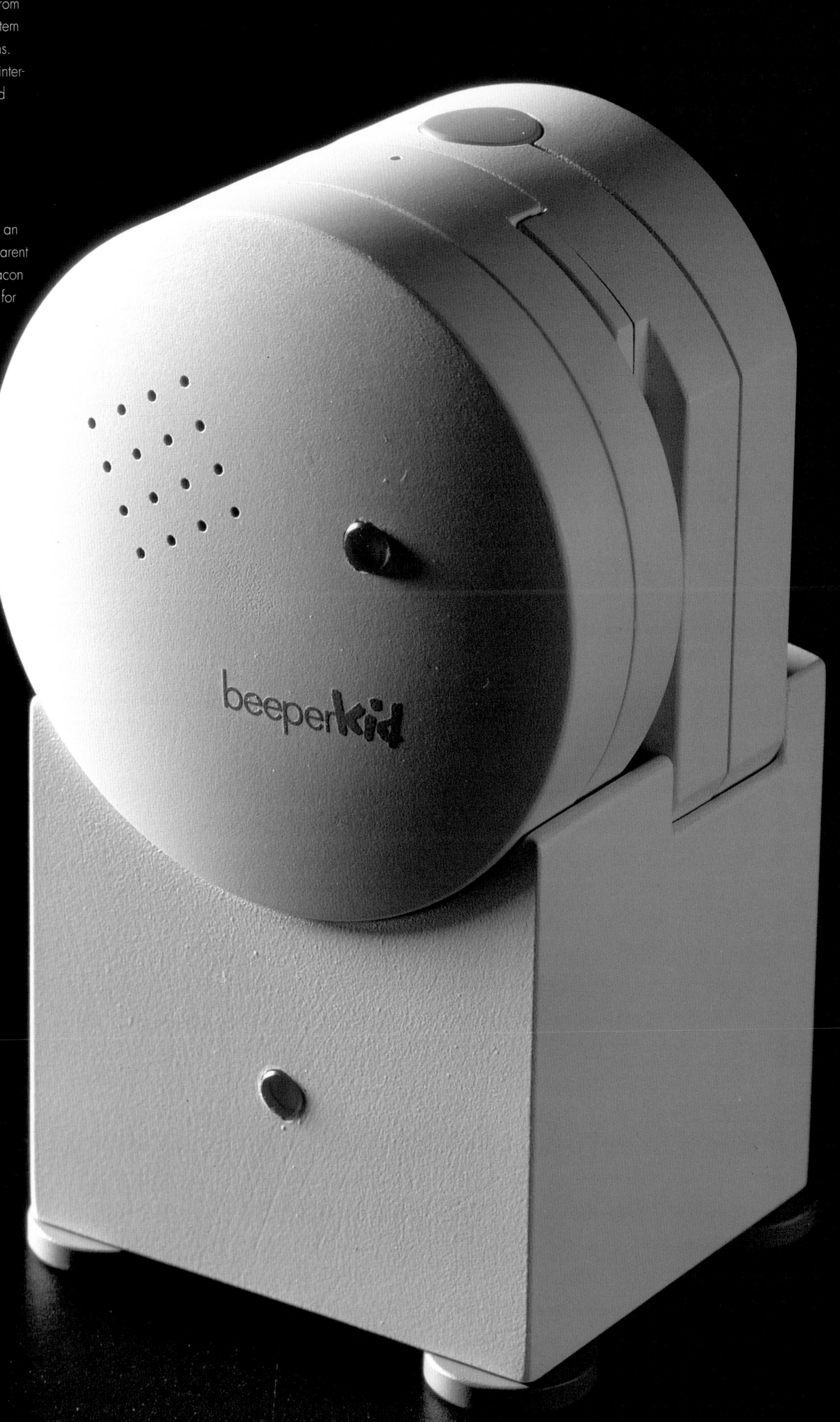

Airbass

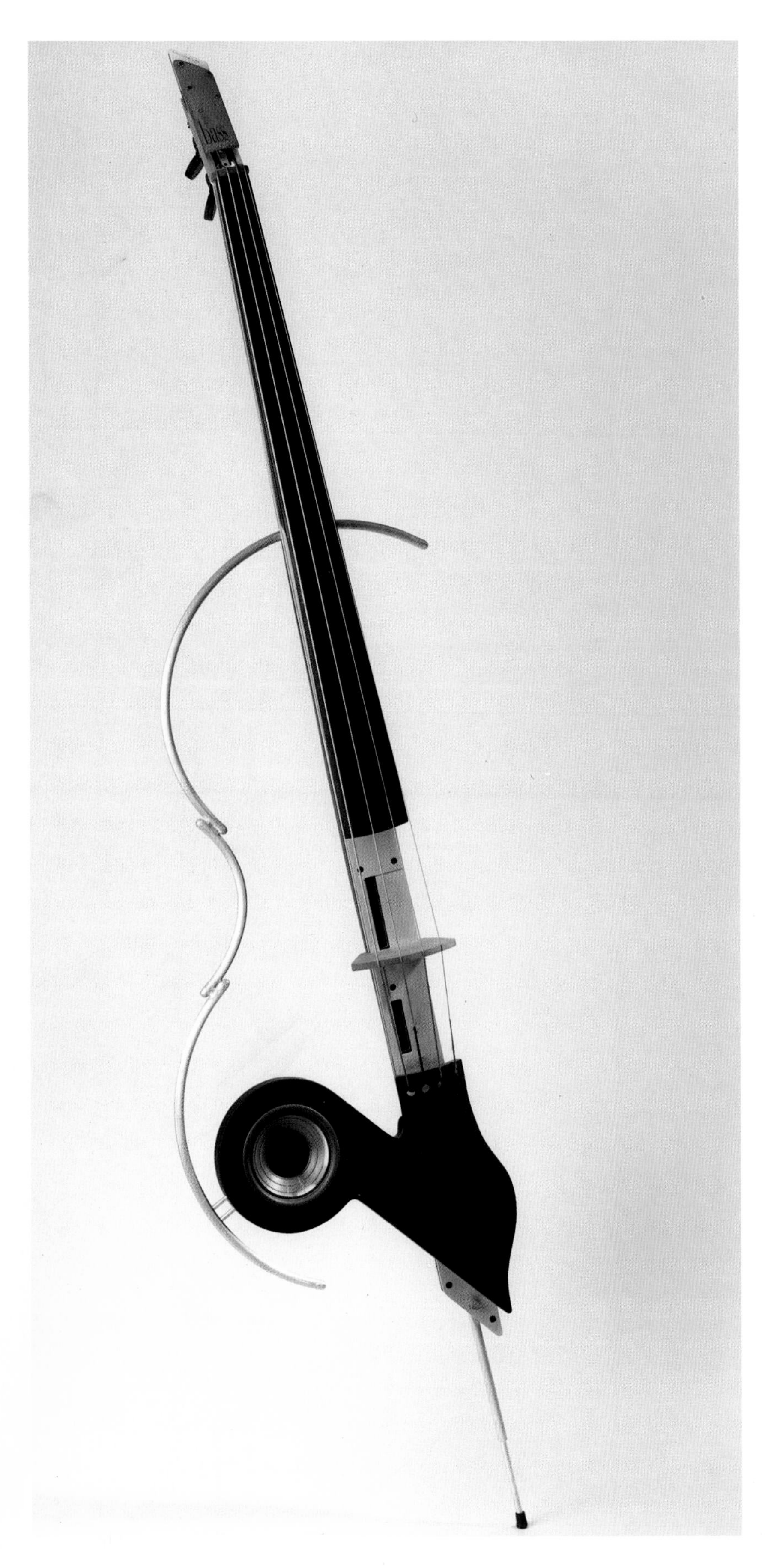

DESIGN OBJECTIVE

To capture the spirit of jazz with freshness and whimsy in a compact, durable, and unintimidating electric bass.

DESIGN SOLUTIONS

An aluminum sandwich core forms the main body to which all else is attached, including an ⅛-inch aluminum spine which travels the length of the strings to support the more than 300 pounds of string tension. A sealed resonating chamber behind the speaker accommodates the natural amplification in an acoustic bass, and a single aluminum spine in the center back of the neck houses the four tuning mechanisms in succession, slenderizing the peg box. Bent aluminum outlines the elegant classical form of the acoustic bass while providing a physical barrier by which the musician can gain a leveraged body position.

RESULT

The Airbass design strips down the standard bass form into a slender, compact electric version, integrating the amplified sound directly into the instrument itself.

DESIGNER
Paolo Lanna

DESIGN SCHOOL
Art Center College of Design

"...Altogether surprising is the successful integration of three divergent design vocabularies: the sculptured shaft of wood and aluminum; the technically minimal housing of the amplifier, set off on an eccentric angle; and the antic outline of a traditional bass."

JURY CHAIR ARNOLD WASSERMAN, IDSA

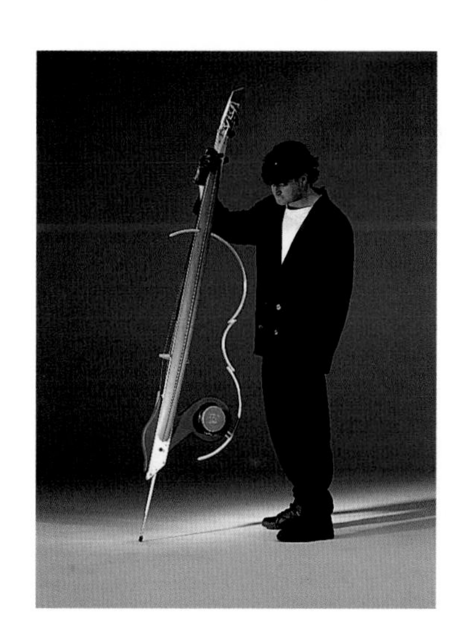

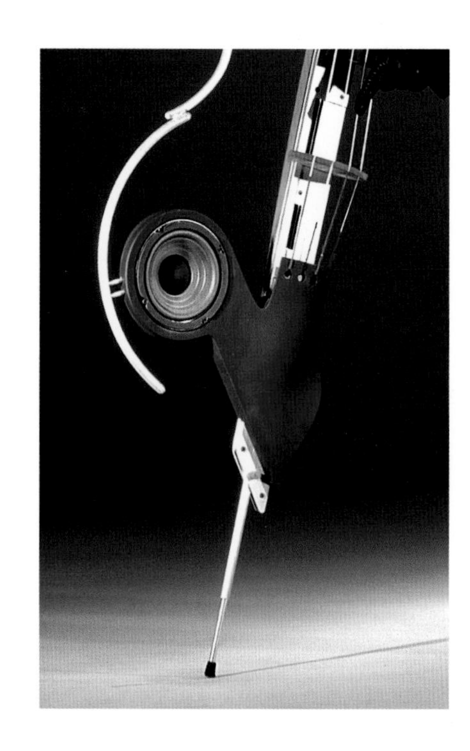

Precedence Bath

"...The warm, smooth relationships of surface configuration, fiberglass materials, colors, surface finishes, smooth, rounded edges and planes, enhance the feel and sense of humanity."

JUROR DAVID JENKINS, IDSA

"Successful design embodies all necessary compromises of functionality and manufacturing to supply the consumer with a pleasing product for the best value."

KOHLER CO.

DESIGN OBJECTIVE

The primary objectives were accessibility and simplicity. The bath had to eliminate barriers to entering and exiting, and be easy to use and easy to understand for all ages and abilities. In style, design, and color, it had to be compatible with other bathing products, fitting into the conventional 5-foot bath alcove.

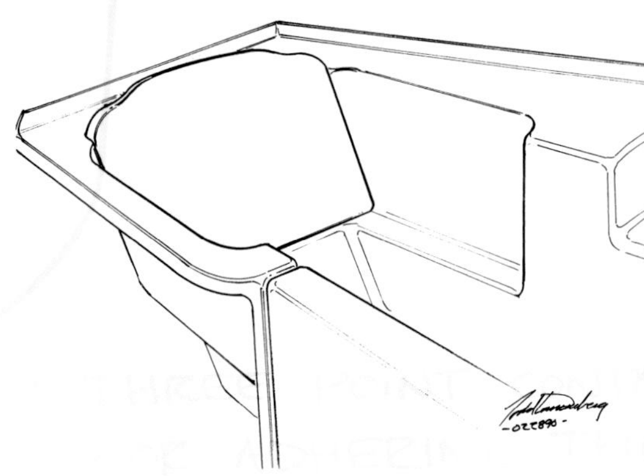

DESIGN SOLUTIONS

In designing the water-tight door, aircraft technology provided the solution with an inflatable seal similar in design to those used on a jet's doors. When two integrated sensors detect water, a continuous duty compressor runs for ten-second cycles to inflate the seal. Then an air tube from the pump to the seal passes through a hollow hinge pin. The door is lowered onto the pin, the air tube is connected to the pump and the seal, and the installation is complete without any hardware or fasteners.

RESULT

This design blends comfort and safe access through the inclusion of an innovative door. Considering that the bath is one of the highest causes of injury in the home, the entry gate and pull-down seat are innovations valuable to everyone.

OTHER AWARDS

–*Interior Design*, ROSCOE 1992

–*Popular Science*, 1993 Design and Engineering Award

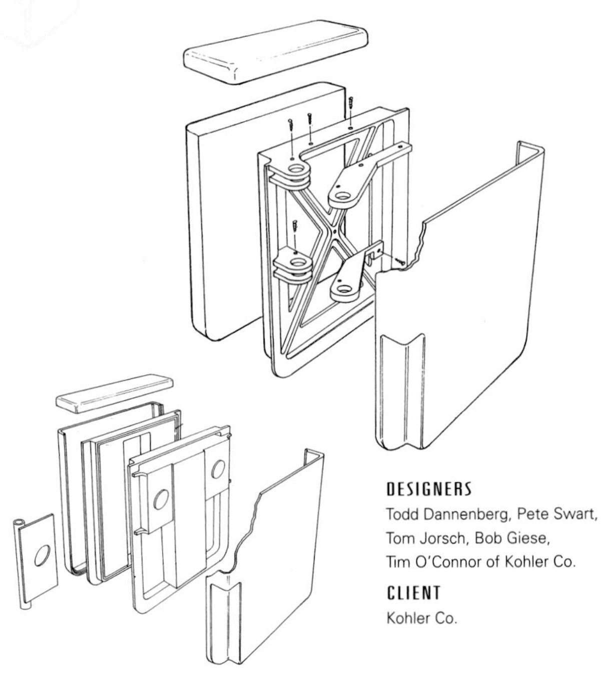

DESIGNERS

Todd Dannenberg, Pete Swart, Tom Jorsch, Bob Giese, Tim O'Connor of Kohler Co.

CLIENT

Kohler Co.

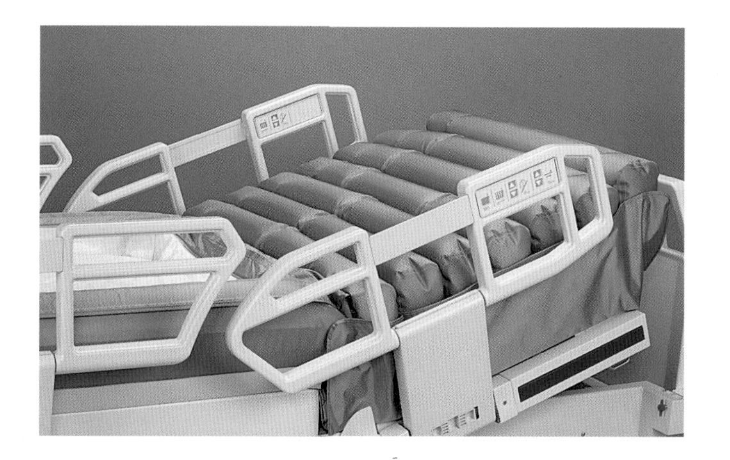

CLINITRON® ELEXIS™ Air Fluidized Therapy

"This unique therapy system combines two entirely different support technologies for patients with advanced pressure ulcers, skin grafts, burns, etc. The challenge to our design team was to visually integrate these two distinct treatment methods in a way that resulted in a professional nonthreatening healing environment."

CHARLES M. HUCK, PRESIDENT OF
HUCK AND STUDER DESIGN, INC.

DESIGN OBJECTIVE

To create a visually integrated unit that combined the functional attributes of two types of patient support surfaces–Air Fluidized Therapy and Low Airloss Therapy–that would also allow patients to adjust their upper body position, and be easier for caregivers to maneuver patients and access controls.

DESIGN SOLUTIONS

From the waist down, Air Fluidized Therapy provides a low pressure, floatation-like patient support with 1,200 pounds of perfectly round ceramic beads set in motion by low pressure, warm dry air beneath a monofilament sheet. From the waist up, Low Airloss Therapy, which uses microprocessor-controlled "smart pillows" filled with air, automatically adjusts pressures relative to the patient's weight and movements. An inflatable bulkhead allows smooth transition between the two surfaces. With no room beneath the support surfaces, a modified parallelogram mechanism allows the side guards to retract upon themselves.

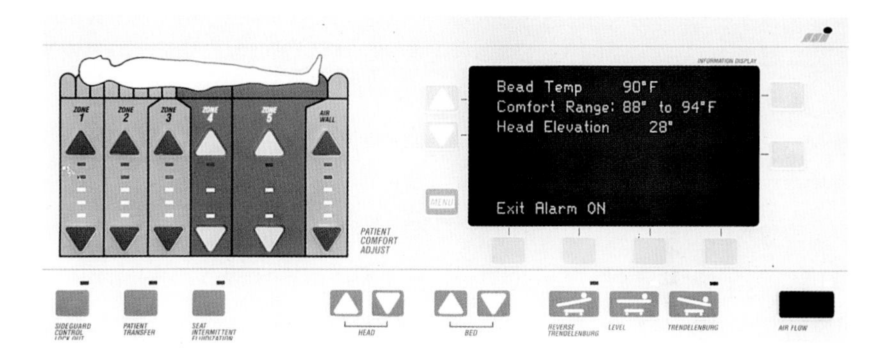

RESULT

The CLINITRON® ELEXIS™ allows patients to lie on their wounds while healing. Oxygen and blood still flow to the wounds–essential to healing–because patient support pressure is drastically reduced by the two therapies and the microprocessor-controlled bed frame.

DESIGNERS

Charles M. Huck, Charles Ashley of Huck and Studer Design, Inc.;
Mark Liebetrau, Sohrab Soltani of Support Systems International, Inc.

CLIENT

Support Systems International, Inc., a wholly owned operating company of Hillenbrand Industries, Inc.

Aquatread

"Once again this year, the Goodyear Aquatread ($85 average price) topped our ratings by a small margin. It performed particularly well in our wet pavement tests, in both hard stops and cornering. It also braked and cornered very well on dry pavement, and our testers liked the way it handled through tight turns."

CONSUMER REPORTS, FEBRUARY 1994

DESIGNERS
Paul B. Maxwell, Sam P. Landers, John Attenello, William Glover, Norm Anderson, Jim Stroble, Don Vera, Joe Hubbell, William Egan of The Goodyear Tire and Rubber Co.

CLIENT
The Goodyear Tire and Rubber Co.

DESIGN OBJECTIVE

To create a tire which would not sacrifice tread wear to gain wet traction.

DESIGN SOLUTIONS

The innovation comes from bridging the tread elements together into a continuous rib that extends from the aquachannel to the shoulder of the tire. Since softer rubber compounds adhere the tread to the road but wear more quickly, a new SIBR synthetic rubber compound was developed to give Aquatread its dual traction/tread wear capabilities.

RESULT

The bridge in the Aquatread contributes to overall stability and handling, and translates into a 60,000-mile tread life warranty. With initial sales exceeding expectations by 50 percent, the tire has significantly shorter stopping distance over conventional broad market, high-mileage tires.

OTHER AWARDS

- *Popular Science,* 1992 Best of What's New
- *Popular Mechanics,* 1992 Design and Engineering Award
- *Advertising Age,* 1992 Marketing 100
- American Marketing Association, 1993 Edison Award
- Intellectual Property Owners, 1993 Distinguished Inventor Award
- American Marketing Association, 1993 Silver EFFIE Award
- Enterprise Development Inc., EDI Innovation Award

DESIGNERS
Doug Birkholz,
Craig Melter, Steve Ruelle,
Paul Hendon of Fiskars Inc.
CLIENT
Fiskars Inc.

Softouch Scissors

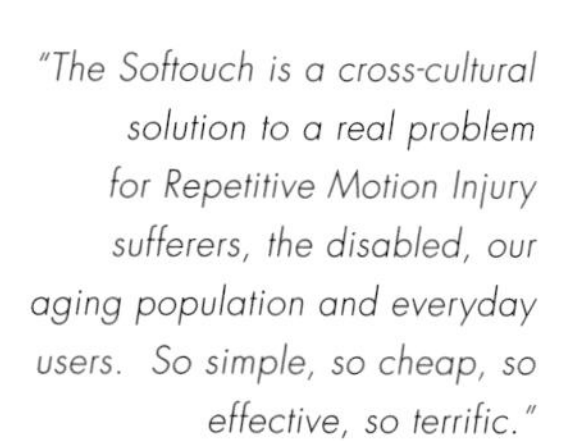

"The Softouch is a cross-cultural solution to a real problem for Repetitive Motion Injury sufferers, the disabled, our aging population and everyday users. So simple, so cheap, so effective, so terrific."

JUROR RALPH OSTERHOUT, IDSA

"...Fiskars is the first manufacturer to cover the handles with a layer of shock absorbing plastic rubber and put it together with other easy-to-use features. These include offset handles with a flat bottom that allows the company's Softouch scissors to slide smoothly across the cutting surface."

THE WALL STREET JOURNAL, MAY 13, 1994

DESIGN OBJECTIVE

To create scissors to be mass produced for the general public while still filling a special niche for people with limited hand use.

DESIGN SOLUTIONS

The Softouch gives people with low hand strength the ability to use scissors. The handles are elliptical in shape when pressed together, big enough for the entire hand, and spring loaded so they don't have to be pulled open after each cut. The long-handled scissors give greater mechanical advantage, and are ambidextrous so the user can alternate hands. Made of soft rubber, the handles absorb shock in the palm and add comfort.

RESULT

Since introduction, the Softouch has crossed into more niche markets than any other scissor in Fiskars' product line.

OTHER AWARDS

–1992 Gov. New Product Award
–Wisconsin Society of Professional Engineers Certificate of Merit
–1992 Mature Market Design Competition, American Society on Aging Finalist, Silver
–*PCM* , 1993 Award of Excellence

Genesis Softside EasyTurn™ Luggage

DESIGNERS
The American Tourister Design Staff
CLIENT
American Tourister

DESIGN OBJECTIVE

To design a truly maneuverable pullman with improved aesthetics and user interaction, and a new wheel system.

DESIGN SOLUTIONS

To achieve the greater stability and mobility desired, the designers drew inspiration from the dollies at the factory, with two fixed center wheels, two rotating end wheels and a rigid pull handle. To withstand rough baggage handling and severe travel conditions, the pullman is made of a new 1680 denier ballistic fabric, a virtually tear-proof blend of nylon and polyester. The handle and wheels are constructed of glass-filled nylon, and the wheel skirts and internal components of high-impact polypropylene.

RESULT

Initial sales figures surpassed forecasts by close to 30 percent. American Tourister's patents on the EasyTurn Tracking System and the rigid pull handle have prevented outright copying by competitors.

OTHER AWARDS

–*Travel & Leisure*, Mark of Innovation Finalist Award
–*I.D.* 1993 Annual Review
–"American Competitiveness" Exhibition selection sponsored by US Commerce

"...The Genesis Softside really is easy turn, rather than easy topple! The design's detailing and ruggedness makes an elegant, appropriately understated accessory that travelers will welcome."

JUROR VINCE FOOTE, FIDSA

THERMOS

DESIGNERS
Greg Breiding, Susan Haller,
Dale Benedict of Fitch, Inc.;
Monte Peterson, Fred Mather of
The Thermos Co.

CLIENT
The Thermos Co.

"...It is small and compact yet looks substantial. In it, Thermos and Fitch went beyond the "zero defect" definition of product quality to include durability, ease of assembly, ease of operation and energy and space efficiency...."

JUROR LOU LENZI, IDSA

"...Fine furniture, as opposed to outdoor equipment or conventional grills, was the metaphor used for the design. The dominant visual element is the double-wall steel dome. The elliptical shape evolved from the need for easy manufacturability...."

DESIGN MANAGEMENT JOURNAL, FALL 1993

DESIGN OBJECTIVE

To create a 110-volt electric grill using a newly developed cooking surface and an insulated, energy-efficient cover that stays cool to the touch.

DESIGN SOLUTIONS

The grill was designed to be assembled in 15 minutes or less. A back-lighted polycarbonate label ensures that the control panel is easy to read at night. The grill's triangular footprint allows it to fit into corners, and provides more usable surface area for cooking.

RESULT

Due to its insulated dome and high-efficiency cooking surface, it uses 50 percent less energy than gas grills. Already, it is the standard for electric grills, opening up a new market segment for Thermos.

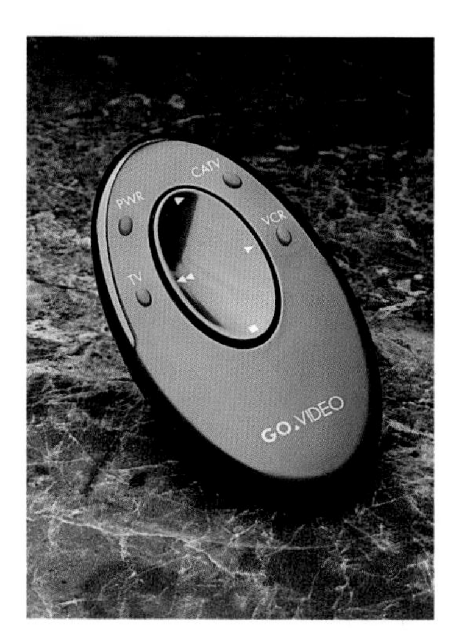

Palm-Mate

DESIGNER
Doug Patton of Patton Design

CLIENT
Go-Video

"... This touchy-feely approach to technology is a welcome departure from the intimidating, hard-edged, techno-look aesthetic typical of most electronics products...."

JUROR LOU LENZI, IDSA

"Even with the $30 Palm-Mate in limited distribution, Go-Video says it can't keep up with the demands. The company sold close to $10,000 in its first 45 days on the market, and in May it ran 300 percent over projected sales."

BUSINESS WEEK, JUNE 7, 1993

DESIGN OBJECTIVE

To design a remote control that put consumer needs on par with manufacturers and dealers.

DESIGN SOLUTIONS

The 35-degree elliptical shape evolved from the natural shape of beach rocks worn by waves and a bar of soap molded by use. A thumb imprint evolved into the navigation pad. Once the thumb is placed on the button, it doesn't have to move, only rock forward, back, left or right. Seldomly used buttons are small, located around the navigation pad.

RESULT

This project has given the designer the motivation to explore the multitudes of shapes and forms that relate to visually exciting and ergonomic products.

Northstar V-8 Engine

DESIGN OBJECTIVE
In addition to performance, basic engine system objectives included quality, reliability and durability, as well as smoothness and quiet operation.

DESIGN SOLUTIONS
The design's unique Exhaust Gas Recirculation system helps reduce oxides of nitrogen. It is designed to be easy to service and, except for fluids and filters, needs no maintenance for 100,000 miles. The die-cast block's design greatly increases the rigidity of the engine, clamps the crankshaft more tightly and reduces noise. The 32-valve configuration is efficient and powerful and the design solution reflects this look.

RESULT
The Northstar has made such a significant impact on Cadillac Motor Division's bottom line that it will ultimately become the only engine under the Cadillac hood.

"Take a highly technical, innovative and complex assembly like an automobile power plant and, rather than hide it away, highlight it as a major sales feature: that's making a statement...."

JUROR CHARLES ALLEN, IDSA

DESIGNERS
Donald J. Schwarz, Thomas Bradley of GM Design Center

CLIENT
Cadillac Motor Car Division

Access

DESIGNER
Cary R. Chow

DESIGN SCHOOL
California State University–Northridge

"...The jury especially admired the fact that the designer has recognized the potential value to the blind of the mobility and mapping guidance which will soon be available in our cars."

JUROR CHARLES BURNETTE, Ph.D.

DESIGN OBJECTIVE

To assist the blind in mobility by integrating several computer-based technologies which would allow a more complete presentation of the surroundings.

DESIGN SOLUTIONS

A compact, tactile display unit, worn on the torso, contains a computerized mapping system that displays a path for the user to follow to his or her destination. The system tracks the user's progress and provides information through the tactile display, vibratory indicators and audio feedback. The design employs a total integration of advanced computer technologies in development or already on the market: global positioning system, shape memory alloy-based interactive tactile display; RISC based microprocessor; erasable optical drive for mapping information; expert systems/artificial intelligence for path creation; and personal information manager/relational database for relational lists/groupings of paths.

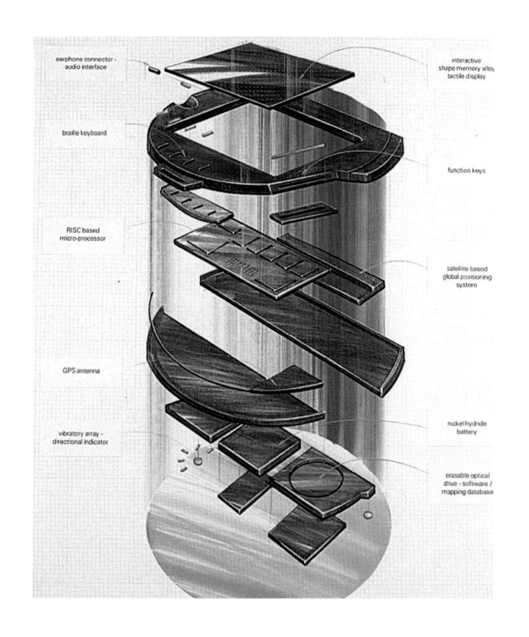

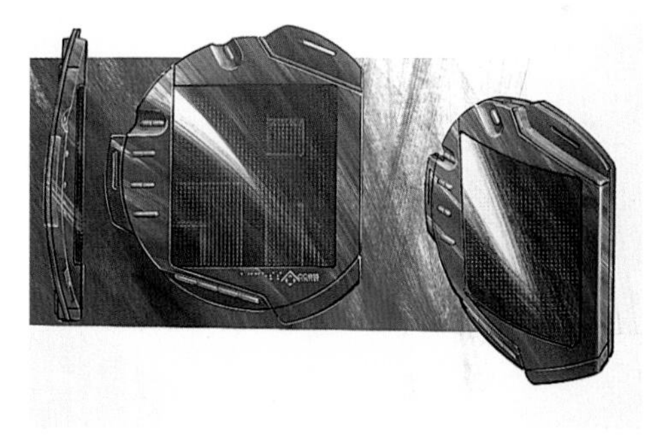

RESULT

The aesthetic of the product evokes images of portable stereo equipment while rejecting the assumption that all products for the blind must be white.

OTHER AWARDS

–Unisys Design Competition, 1991 Second Place

Deskjet Portable Printer

"...Jim Girard clearly gave careful thought to a number of issues, such as offering a range of sheet feeding options, and then did a great job of organizing the different elements."

JUROR CHIPP WALTERS, IDSA

"Design illuminates rather than obfuscates the use of man-made tools and products. Successful design invites the user to explore and interact with it, yet it should be elegant and beautiful to look at."

HEWLETT-PACKARD

DESIGN OBJECTIVE

To create the smallest, lightest, and most flexible portable printer possible.

DESIGN SOLUTIONS

A totally self-contained printer can be "plugged" into a separate sheet feeder allowing the printer to perform both portable and desktop printing needs. The "V" design made the printer's control panel and sheet feeder functions easy to view and access, and created the smallest possible footprint on the desktop surface. When folded flat, the product is easy to transport.

RESULT

This fully featured, high-performance printer weighs only 4.4 pounds and sets up with one hand. Environmentally sound, the inkjet cartridges can be mailed back to Hewlett-Packard for recycling.

DESIGNER
Jim Girard of Hewlett-Packard
CLIENT
Hewlett-Packard,
Asian Peripheral Div.,
Republic of Singapore

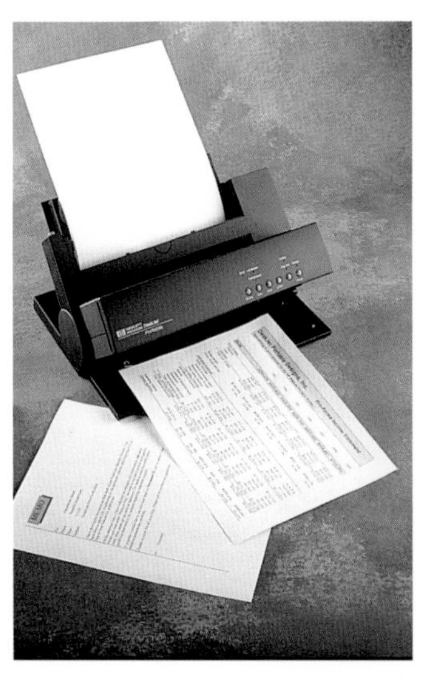

Airflex™

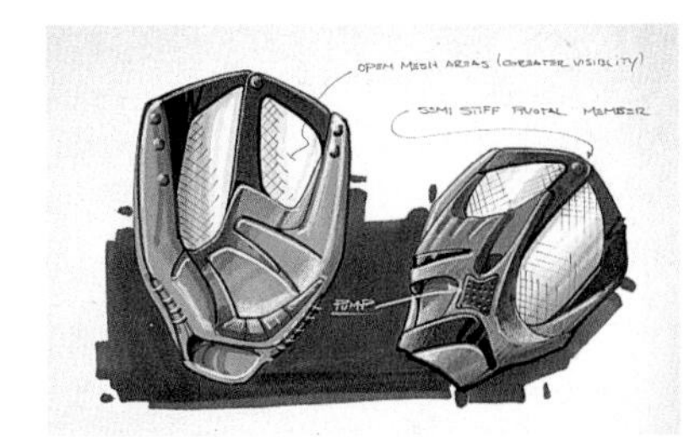

DESIGN OBJECTIVE

To design a glove which would provide a more customized fit and reduce "break-in" time, while still retaining the form, feel, and look of a traditional glove.

DESIGN SOLUTIONS

An air bladder system in the back of the glove with an inflation bulb and release button on the thumb allows each user to tailor the glove's fit. They are easily accessible to the user's ungloved hand, and are not in the way when equipment pounds and softens the leather at the end of the manufacturing process. Flex points, hinges, and stretch points on the glove allow a wider range of hand motion. Lycra and neoprene used in small quantities improve the glove's flexibility.

RESULT

The Airflex design, which was based on an analysis of hand motion, finger force and glove closing dynamics while catching a ball, achieves optimal functional performance.

"Successful design is based on innovation and tradition. Here, we integrated a revolutionary technology into a historically traditional product.... By keeping leather and stitching details of the traditional glove, the new design maintains traditions linked to America's traditional game."

DESIGN CONTINUUM INC.

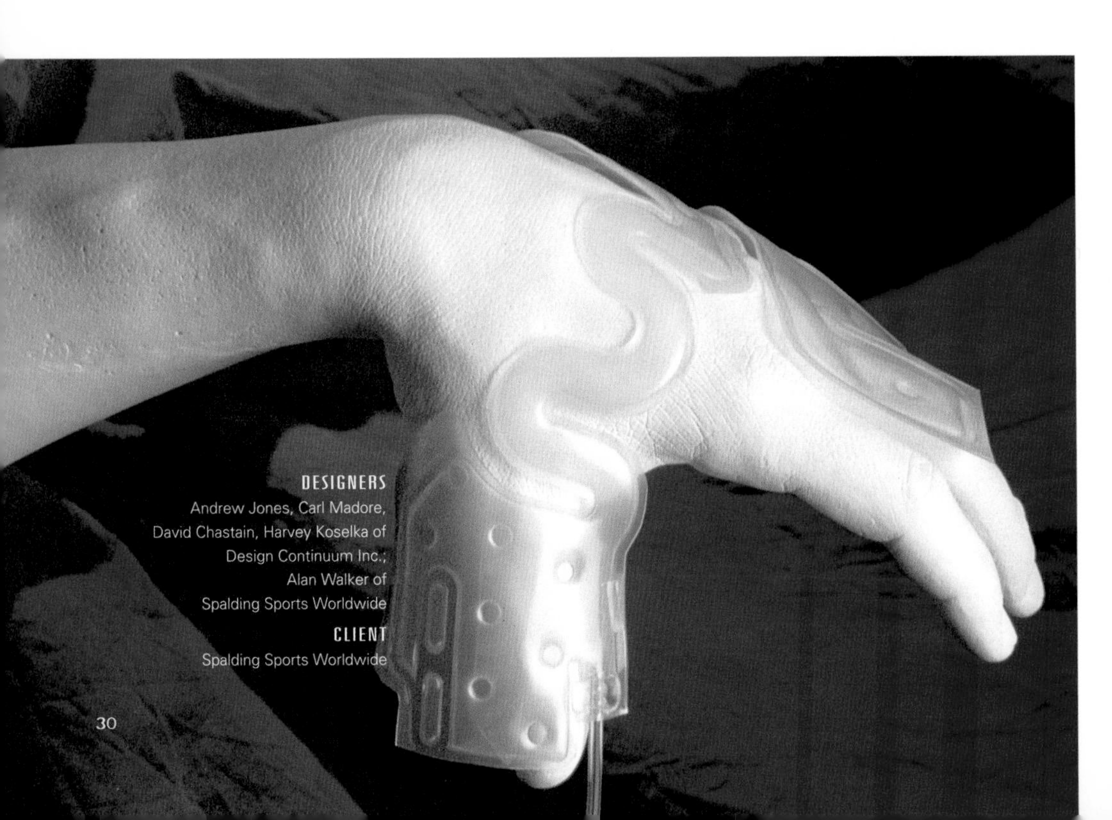

DESIGNERS
Andrew Jones, Carl Madore, David Chastain, Harvey Koselka of Design Continuum Inc.; Alan Walker of Spalding Sports Worldwide

CLIENT
Spalding Sports Worldwide

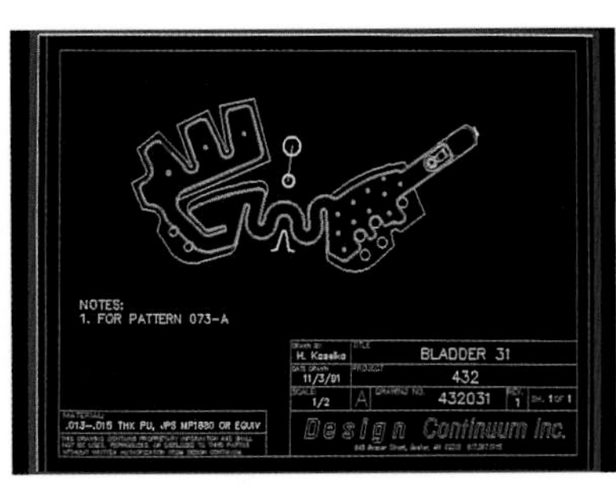

Tadpole Infant Positioning System

DESIGN OBJECTIVE

To design a product for infants that would optimize physical therapy in a home setting, be nonthreatening to parents, and noninvasive to their environment.

DESIGN SOLUTIONS

The multiple positions used in therapy–prone, supine, side laying and sitting–required a system of shapes that were self-descriptive in design so they could be easily assembled and rearranged. They are made of soft molded foam covered with a non-toxic, flexible urethane elastomer that is seamless and waterproof, making them easy to clean and durable.

RESULT

The Tadpole has met with extraordinary market success. In fact, it has exceeded the first year sales forecast by 35 percent. Numerous requests have been made for a larger version of the Tadpole that can be used with older children.

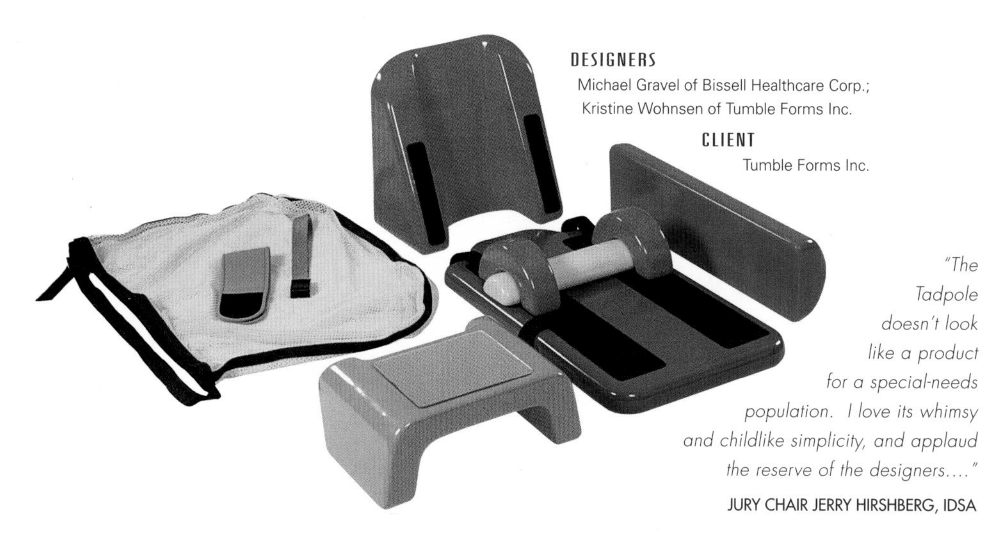

DESIGNERS
Michael Gravel of Bissell Healthcare Corp.;
Kristine Wohnsen of Tumble Forms Inc.

CLIENT
Tumble Forms Inc.

"The Tadpole doesn't look like a product for a special-needs population. I love its whimsy and childlike simplicity, and applaud the reserve of the designers...."

JURY CHAIR JERRY HIRSHBERG, IDSA

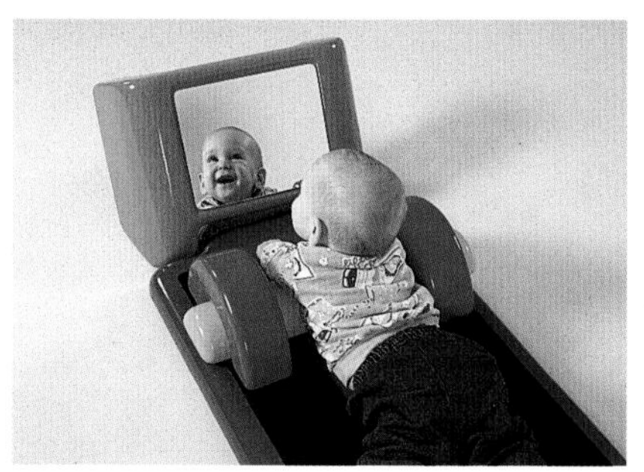

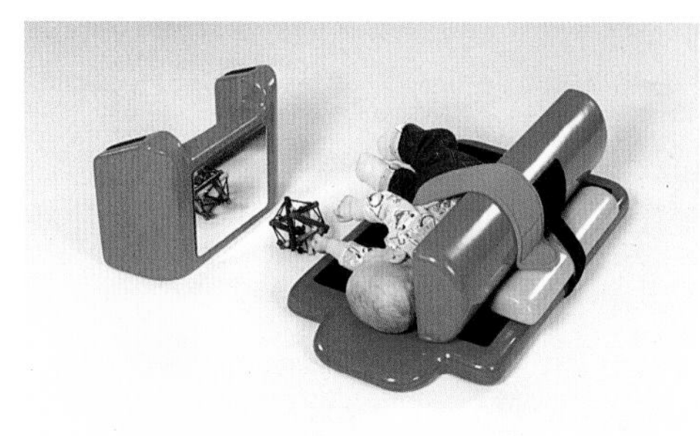

FM 560 Motorized Stairclimber

"An elegant design departure with simple controls, solid performance, small footprint and modest price make the FM 560 Stairclimber hard to resist...."

JUROR RALPH OSTERHOUT, IDSA

DESIGN OBJECTIVE

To provide the smooth, addictive feel and effective performance of a health club climber at only a fraction of the size and cost for home use.

DESIGN SOLUTIONS

The base folds for convenient storage, while allowing the stairclimber to be shipped fully assembled. Extensive use of engineering plastics in major structural applications combined with a strict form meeting function requirement for every component kept costs down. For instance, the triangular side plates that cover the motor also serve the structural purpose of helping to hold the unit up.

RESULT

At a cost one-fourth the price of a health club climber, the FM 560 offers users a choice of three preprogrammed workout patterns and the unique surround handlebar facilitates proper posture.

DESIGNERS
Jim Easley, Rick Polk, Doug Soller,
Ken Schomburg of Leisure Design Assoc.;
Hans Friedebach of Fitness Master, Inc.

CLIENT
Fitness Master, Inc.

Handkerchief TV

DESIGN OBJECTIVE

Emilio Ambasz Design Group envisioned the elimination of the hard plastic case that always encloses our electronic equipment.

DESIGN SOLUTIONS

This television is made like a handkerchief folded into four, with a 3- or 4-inch flat screen on one quarter, the speaker and antenna on another, and the rechargeable battery and other equipment on the fourth. "The technology to create such a product already exists," asserts Emilio Ambasz, "All we need is a different way of perceiving the housing of products."

RESULT

A portable, flat-screen TV set in a leather handkerchief that fits in a shirt pocket when folded up.

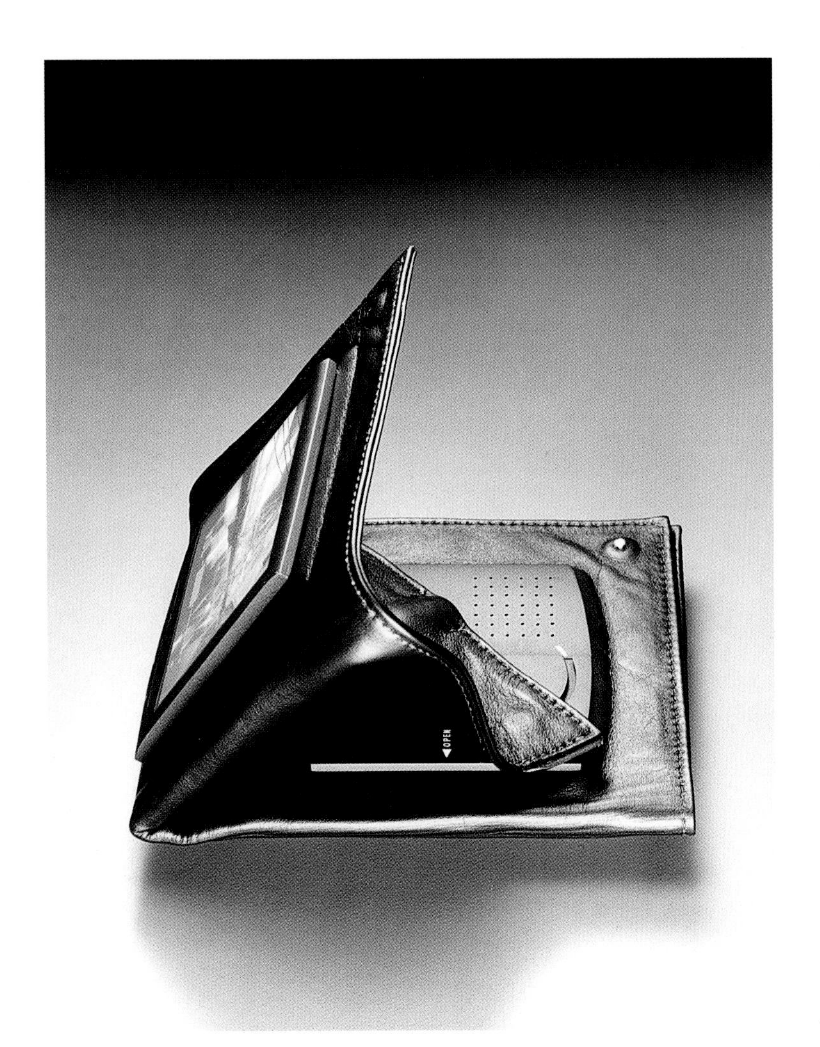

"...Emilio Ambasz has transformed television–a technology, process and product–into a soft, personal, individual, lovable, comfortable and strokable object. Soft TV. It's the puppy-fication of television."

JUROR RITASUE SIEGEL, IDSA

DESIGNERS
Emilio Ambasz,
Masamichi Udagawa of
Emilio Ambasz Design Group
CLIENT
Brion Vega

Bunjieboarding—Bunjie System

DESIGNERS
George O. Podd, Greg Holderfield of George Podd Design

CLIENT
George Podd Design

DESIGN OBJECTIVE

The Bunjie System grew out of a long recreational history on the Midwestern flatlands where the lack of surfable waves or skiable mountains compels youths to create their own forms of thrill-filled momentum.

DESIGN SOLUTIONS

A flexible cord, a pivoting device, and a handle are the three basic components that comprise the Bunjie System. The pivot is affixed to the ground, and attached to it is a 10-foot length of rubber cord with a handle at the opposite end. Slingshot acceleration, combined with a skater's innovative technique, provides for unlimited movement within the full stretch radius of the rubber cord.

RESULT

The design slingshots the skateboarder around a central axis, making it possible to skateboard indoors. The system allows for endless hours of skating in a designated area, luring skaters away from streets.

"The designers had a kinetic idea that they built and had fun using. Looking at this design is like seeing the invention of baseball."

JUROR TUCKER VIEMEISTER, IDSA

Animal

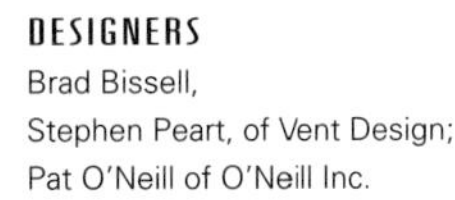

DESIGNERS
Brad Bissell,
Stephen Peart, of Vent Design;
Pat O'Neill of O'Neill Inc.

CLIENT
O'Neill Inc.

"...Bigger is not better in sports equipment, so the designers created a way to make the warmer wet suit more flexible by removing material—demonstrating once again that less is more."

JUROR TUCKER VIEMEISTER, IDSA

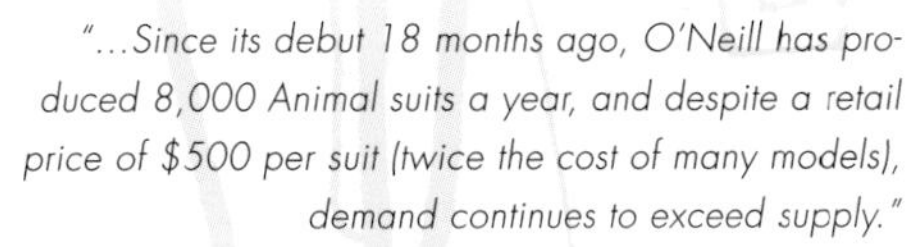

"...Since its debut 18 months ago, O'Neill has produced 8,000 Animal suits a year, and despite a retail price of $500 per suit (twice the cost of many models), demand continues to exceed supply."

I.D., JANUARY/FEBRUARY 1992

DESIGN OBJECTIVE

To create a wet suit that efficiently retains body heat and accommodates user movement with minimum restraint.

DESIGN SOLUTIONS

Retaining body heat dictates that the neoprene material be as thick as possible, but freedom of movement requires a highly elastic suit—thin neoprene material. Molded grooves in the neoprene, oriented anatomically, provide selectively enhanced elasticity in the direction of maximum body movement without sacrificing body heat. Since cutting the grooves with a carbon dioxide laser created a messy, hard-to-remove residue and left a layer of open neoprene cells which retained rather than shed water, molding proved to be the best method, giving a smooth, unbroken surface of closed neoprene cells.

RESULT

The design combines a fully integrated system of molded parts to create the most elastic cold water wet suit on the market. It offers the warmth of a 3mm-density wet suit with the elasticity of a 2mm suit. It has 27 percent more stretch and 20 percent less weight than any full wet suit.

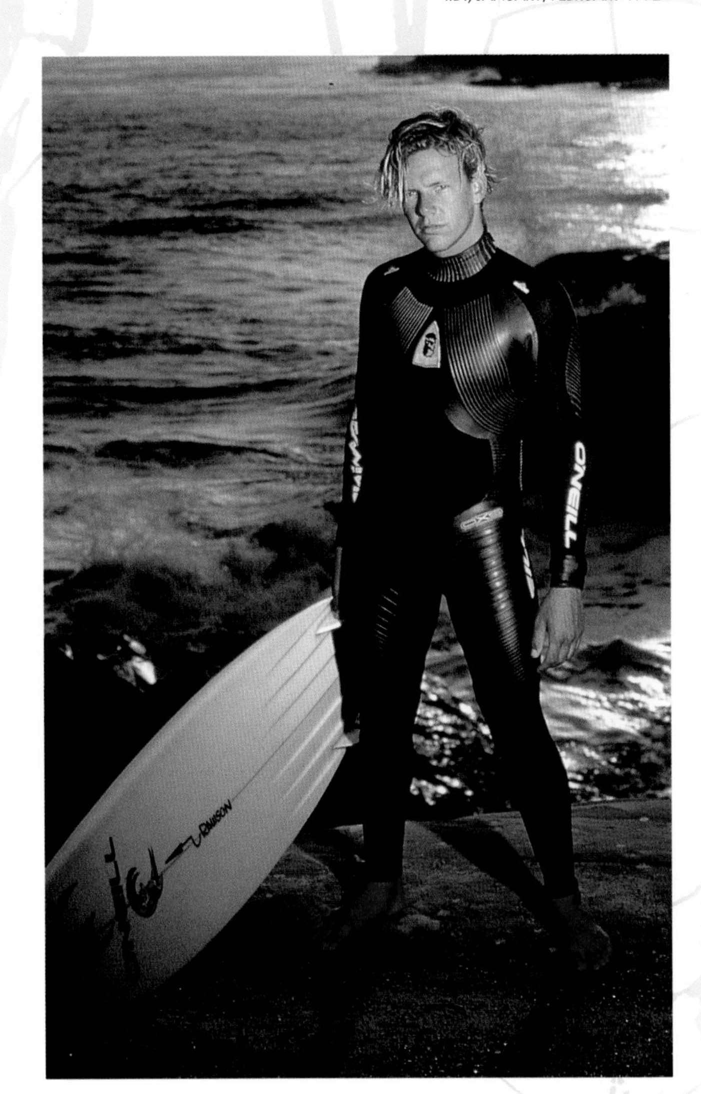

Proformix Keyboarding System

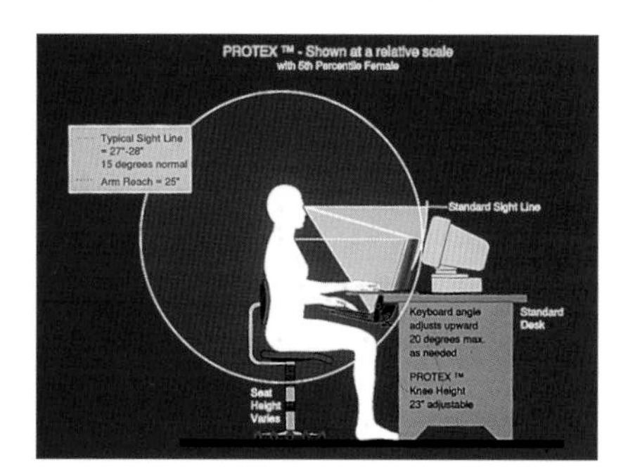

DESIGN OBJECTIVE

To design a system which would significantly reduce repetitive strain motions and cumulative trauma risk factors to the computer user as well as mitigate electromagnetic field exposure; integrate all computer work while accommodating different sized people; and generate an identifiable but understated product appearance that would fit many workstations.

DESIGN SOLUTIONS

The height- and angle-adjustable keyboard is sloped downwards, away from the operator, to reduce wrist strain. Plastic was chosen as the ideal material for the palm rest because it does not draw the user's body heat or transmit electromagnetic fields from the monitor to the operator. To minimize the environmental impact, all the plastic was colored using an EPA-approved colorant free of heavy metals.

RESULT

The design integrates computer peripherals, holding them in adjustable positions that can be easily changed to help reduce the risk of Cumulative Trauma Disorders, the leading cause of occupational illness in the US. The product comes in two versions, both retailing for under $300.

DESIGNERS
Eugene A. Helmetsie,
Alan Hedge, Ph.D., Ben Sherman,
David McClelland of Pelican Design;
Michael G. Martin of Proformix Inc.
CLIENT
Proformix Inc.

"...Proformix goes a long way to correcting ergonomic sins in offices where the budget does not allow new ergonomically correct systems."

JUROR DAVID JENKINS, IDSA

"...'The Proformix Keyboarding System significantly reduced postural risks, eliminated postural discomfort and improved reported productivity,' reports Dr. Hedge. 'Musculoskeletal discomfort in the right wrist of users dropped from 63 to 16 percent.'..."

RISK MANAGEMENT, MARCH 1994

PowerBook

DESIGNERS
Robert Brunner, Gavin Ivester, Susanne Pierce,
Jim Halicho, Eric Takahashi of Apple Computer, Inc.;
Michael Antonczak of In Design;
Matt Barthelemy of Lunar Design

CLIENT
Apple Computer, Inc.

DESIGN OBJECTIVE

To create a small notebook computer with a comfortable-to-use trackball, built-in floppy drive, simply shaped printed circuit board, and easy battery removal.

DESIGN SOLUTIONS

The front-and-center trackball arrangement accommodates both left- and right-handed users. Placing the battery and hard disk drive on either side of the trackball created large, comfortable palm rests. The display's locking latch was positioned in the front of the display so that it could be operated with one hand.

RESULT

By delivering excellent ergonomics in a very compact package, the PowerBook has earned Apple the top market share position. The comfortable palm rests reduce the user's lateral wrist angle when working in a cramped space.

"...It's immediately apparent that the user came first in this design. This is not just an exercise in compacting technology or creating visual appeal, but a rational design that meets a user's need for individual comfort and durability."

JUROR JAMES BLECK, IDSA

PowerBook Duo System

DESIGN OBJECTIVE

To create a feature set of the smallest and lightest components with effortless networking capabilities.

DESIGN SOLUTIONS

A much smaller package was achieved by moving most of the connectors, including SCSI, sound and video, off the main PC board onto a modular, snap-on connector bar. With no cables used to dock the notebook, portability is as easy as pushing a button. To reduce the weight and thickness of the LCD display, the metal frame present beneath the plastic housing in earlier modules was eliminated.

RESULT

A very thin, lightweight, high-performance notebook computer, the Duo can attach to a docking device to provide uncompromised desktop computer functionality. Designed to cost less than two computers, it has outsold marketing forecasts.

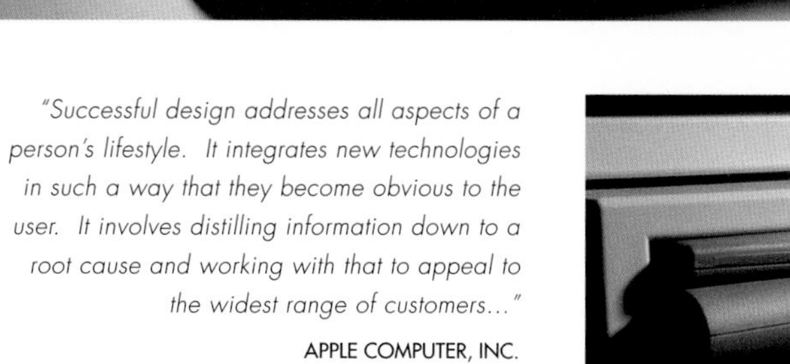

"Successful design addresses all aspects of a person's lifestyle. It integrates new technologies in such a way that they become obvious to the user. It involves distilling information down to a root cause and working with that to appeal to the widest range of customers..."

APPLE COMPUTER, INC.

OTHER AWARDS

–*BYTE* Magazine, 1992 Award of Distinction
–Comdex/Fall 1992, Best Portable System
–*MacUser*, 1992 Editors Choice Award, Breakthrough Product of the Year
–*Personal Computer World, U.K.*, Best Power Portable of 1992

DESIGNERS

Apple Industrial Design Group;
Apple Portable Product Design;
IDEO Product Development;
Lunar Design

CLIENT

Apple Computer, Inc.

Dodge Neon

DESIGNERS
Chrysler Design Staff

CLIENT
Chrysler Corp.

DESIGN OBJECTIVE

To develop a concept car that would propose solutions for all of the changes coming in the late '90s–an economical to build, environmentally friendly vehicle that meets tough CAFE fuel economy requirements and emission standards.

DESIGN SOLUTIONS

The Neon has no "B" pillar on the sides and the doors slide apart, making entry and egress easier. Coded for easy processing, the interior and exterior panels are recyclable, and the wheels are made of recycled aluminum cans. Light carbon fiber composite seat frames reduce weight, improving fuel economy. The Neon is painted with water-borne paints, and features a two-cycle, lightweight, nonpolluting engine which saves fuel and runs clean.

RESULT

The Neon's engine is half the size of a conventional engine and more than 40 percent lighter, providing the Neon the interior and luggage space of a Dodge Spirit in a "footprint" smaller than a Dodge Shadow.

"This design is refreshing–it moves you without relying on traditional symbols of automotive appeal. Its level of design content and thoroughness is unusual for a show car and it should be taken seriously by the manufacturer."

JURY CHAIR JERRY HIRSHBERG, IDSA

My First Sony

Electronic Sketch Pad and Animation Computer

DESIGNER
Sony Design Center

CLIENT
Sony Corp.

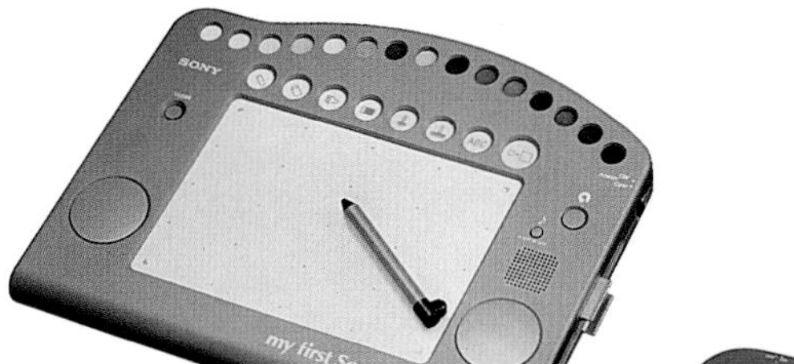

"The exquisitely intuitive design of their interface compels even the most hesitant child to create on these technological bridges between the whimsical art of childhood and the computer-driven world of adulthood."

JUROR RALPH OSTERHOUT, IDSA

DESIGN OBJECTIVE

To create something that allows children to express their creative selves, a modern alternative to the pencil and crayon.

DESIGN SOLUTIONS

Each product takes a fundamentally different approach while using the same medium–video-generated images–viewed on a TV monitor. The Electronic Sketch Pad is based on traditional line drawing and simple coloring techniques, with an artist's palette as the basis for the design concept. The Animation Computer compiles the art of sculpture, collage and painting by combining different shapes to compose a drawing.

RESULT

The great number of different images that can be created using the menu of shapes continues to amaze the design team. Children are using these tools to create things far beyond the imagination of an adult.

Rear Projection TV

DESIGNERS
Richard Bourgerie,
Lou Lenzi of
Thomson Consumer Electronics

CLIENT
Thomson Consumer Electronics

"...The designer has accented the projection screen as a work of art. The simplicity and elegance of the elements surrounding the screen, as well as the reduction of the back cabinet, make the product feel light."

JUROR LIZ POWELL, IDSA

DESIGN OBJECTIVE

To create a compact, freestanding projection cabinet which could be built-in.

DESIGN SOLUTIONS

To emphasize the screen and minimize the frame size, the team designed an aluminum screen frame. A new ported speaker system and crossover allows the audio system to be mounted mid-screen–improving stereo imaging and making the unit easier to build into a wall.

RESULT

This design rethinks the typical box-shaped, rear projection TV cabinet. The result is a unique, much more compact, cost-effective solution that can be built into homes so only the screen and speakers are exposed. The cabinet costs 30 percent less to produce than its predecessor and fits through a 28-inch door opening–an installation benefit. It has contributed significantly to the RCA brand's 51 percent growth in projection TV market share for 1992.

OTHER AWARDS

–Innovations '92, The 1992 International Summer
–Consumer Electronics Show, Design and Engineering Award

Highlander Concept Vehicle

"...the Highlander's mission was to attract attention and future truck customers while testing market features for the client, Chevrolet."

GENERAL MOTORS DESIGN CENTER

DESIGNERS
Donald J. Schwarz,
William A. Davis of
GM Design Center

CLIENT
Chevrolet Motor Division,
General Motors Corp.

DESIGN OBJECTIVE

To create a show vehicle with a special feature that would capture the imagination while attracting attention and future customers.

DESIGN SOLUTIONS

Bright, playful colors–lime green and purple–were chosen to appeal to young buyers and women, now one of the largest growing segments of the truck market. Features include lower ground effects, big tires, a fold out tool box, a back cab panel TV set, and the ability for the back cab panel to fold down and the back window to slide down. Because a concept vehicle requires fabrication without full production tooling capabilities, a series of Buick hood center moldings laced with rib stop nylon formed the hinging device for the roll top desk-type cover.

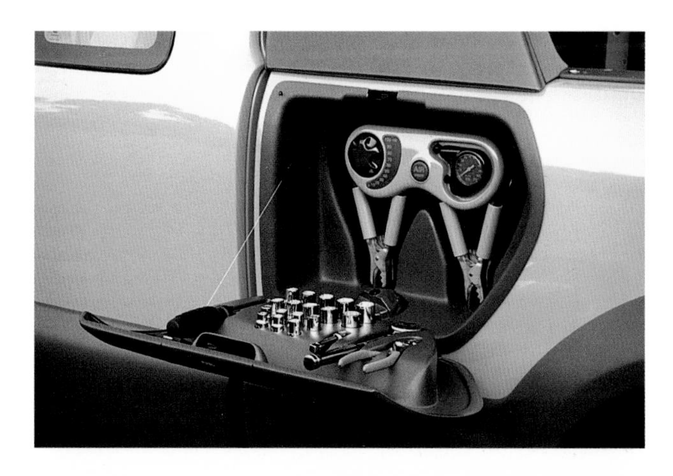

RESULT

The Highlander gave Chevrolet an opportunity to test market features; the rear quarter sliding door will be the most significant as an industry spin-off since it was very well received.

Single-Burner Portable Cooker

DESIGNERS
Young S. Kim; Tony Indindoli of
INNO DESIGN, INC.

CLIENT
Tong Yang Magic Corp. (TYM)

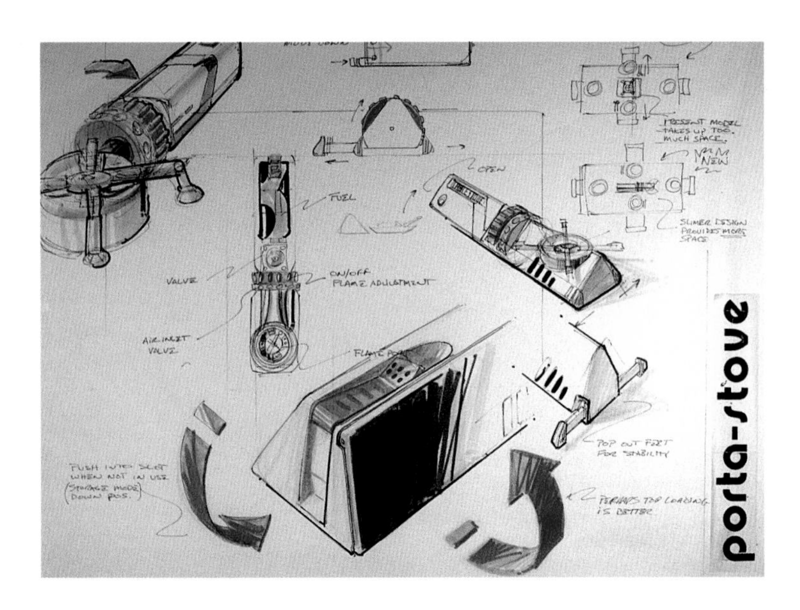

"...Its lady bug imagery of colors and spots is accented by the clever integration of folding 'legs' for compactness. Here nature and technology are fused for the benefit and pleasure of the user."

JUROR RALPH OSTERHOUT, IDSA

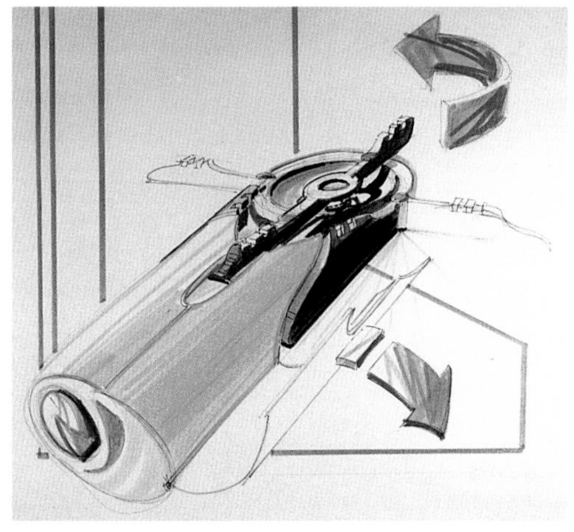

DESIGN OBJECTIVE

To completely re-state TYM's line of single-burner portable cookers with a revolutionary design.

DESIGN SOLUTIONS

The body style and retractable leg configuration forms the backbone of the design. The unit is placed on a flat surface, then the legs are extended from the base and fanned out into a three-point set configuration. The base contains a standard butane tank.

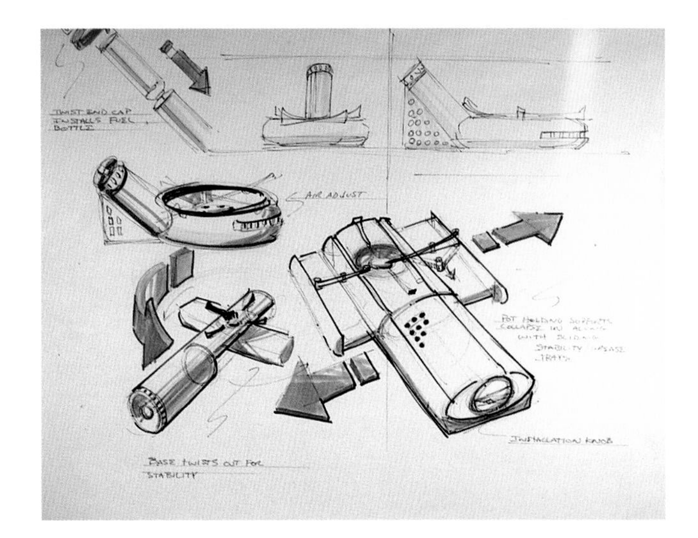

RESULT

The design cuts the footprint of the product by over 60 percent while retaining the needed stability for handling pans. The cooker's visual impact is all the more incredible in light of the commonality of its component parts.

Orchestra Lamp

DESIGNERS
John Rizzi;
Brooks Rorke of RorkeDesign;
Ralph Osterhout of Team Machina;
Andrew Cogan, Karl Konrad of Knoll

CLIENT
The Knoll Group

"...I am especially impressed by the way it combines the most simply activated height adjustment for the light source with an exquisite composition of shade, diffuser and stand...."

JUROR DAVID JENKINS, IDSA

SwingArm Desk Lamp Prototype

DESIGN OBJECTIVE

To design a simple lighting system which could fit well in any office setting and be personalized and customized around working habits.

DESIGN SOLUTIONS

John Rizzi and Brooks Rorke designed a lightweight plastic diffuser to comply with new UL requirements. A friction collet around the shaft makes the lamp's height adjustable without any fasteners or knobs.

RESULT

The lamp features a small footprint, and a telescoping post which makes it easy to tailor its fit on desks under overhead cabinets.

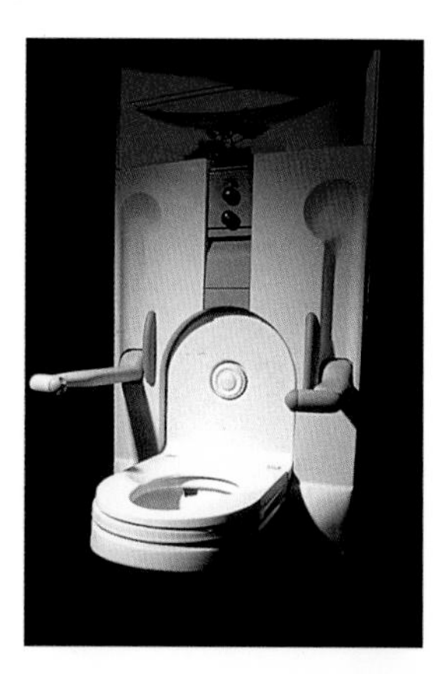

Metaform Personal Hygiene System

"...the designers have provided an environment and fixtures that demonstrate sensitivity to design theory, human factors, technology and materials and puts this knowledge to use in the service of real people...."

JUROR VINCE FOOTE, FIDSA

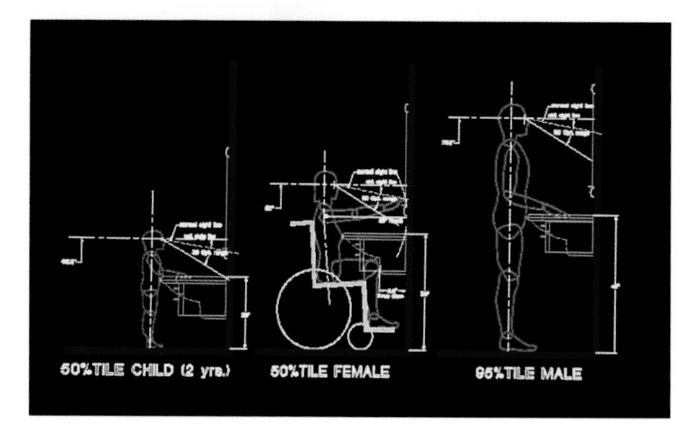

DESIGN OBJECTIVE

The Metaform project's objective was to focus on personal hygiene, allowing disabled and elderly people to live independently and with greater dignity.

DESIGN SOLUTIONS

The Metaform is a modular system of components which can be used individually or as an integrated system. Each module is described as a node of related activities, designed to remove hazards, and adjust easily between different users. For instance, the sink node raises and lowers at the push of a button to accommodate various heights, and the shower node includes a modular support bar/accessory rail that can be installed over any existing stud wall construction. The toilet node adjusts quickly to meet individual needs, and can be automatically washed and sanitized when rotated into the wall for storage.

RESULT

Herman Miller is currently looking for the appropriate manner to commercialize the Metaform system.

DESIGNERS
Gianfranco Zaccai, Timothy Dearborn, Arthur Rousmaniere, Andrew Ziegler, Lynn Noble of Design Continuum Inc.

CLIENT
Herman Miller Research Corp.

Access Power•Pointer

DESIGNER
Vincent Haley

DESIGN SCHOOL
North Carolina State University

SPONSOR
IBM Design Center

"...By using an electrical impulse generated by a puff of air to replace muscle power, the design reduces fatigue and stress on the head, neck and mouth muscles, the only ones now available to this user group...."

JUROR CHARLES BURNETTE, Ph.D.

DESIGN OBJECTIVE

To develop peripheral products for the computer environment that would make existing PCs more accessible to individuals with varying disabilities.

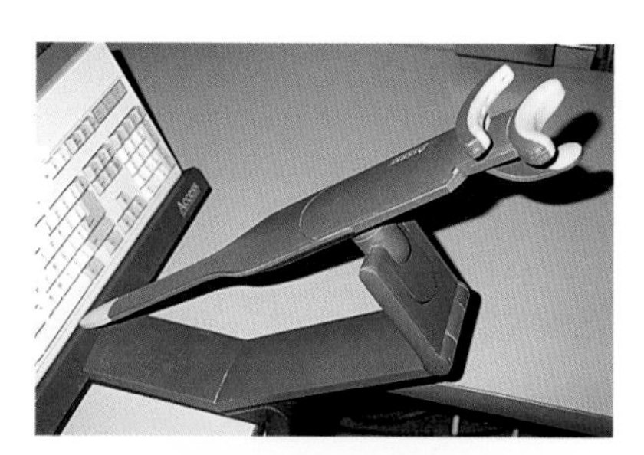

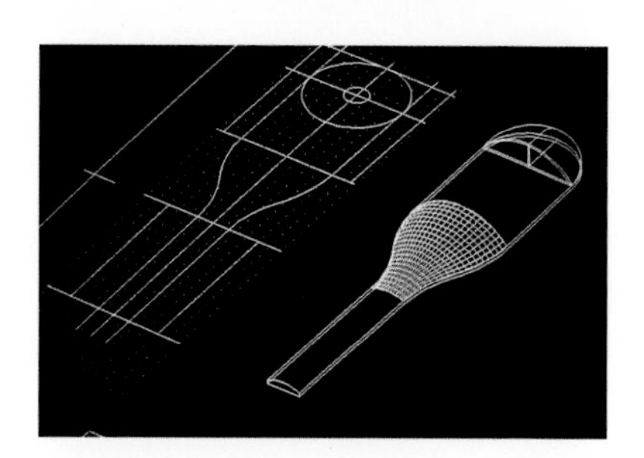

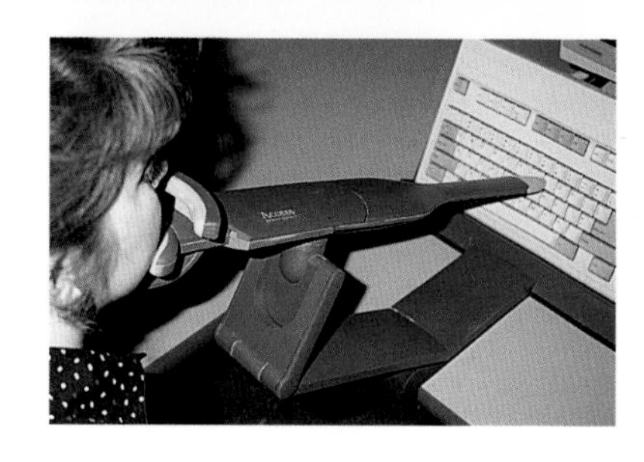

DESIGN SOLUTIONS

The Access Power•Pointer is a chin-controlled input device designed to make PCs accessible to individuals with spinal cord injuries who can only move their head and neck. Positioned directly in front of a PC monitor, the device extends out over the front edge of a work surface. With chin placed in the provided rest, the user simply pivots his/her head, making the pointer follow. When the tip is in position, the user blows lightly. That puff of air converts to an electrical impulse which activates the powered tip, extending it 1½ inches to depress the desired key.

RESULT

The Access Power•Pointer is flexible, and less limiting than traditional adaptive devices. Positioning it on the work surface, with no part of the unit placed in the user's mouth makes this product accessible to a large number of users in an office.

OTHER AWARDS

–Johns Hopkins National Search for Computing to Assist Persons with Disabilities, Regional Winner

New Move Wheelchair

DESIGN OBJECTIVE

To design a wheelchair that was faster, easier to use, and easier to stop. Additionally, the wheelchair would appeal to people whose lack of hand coordination prevents them from being able to rotate a supporting wheel.

DESIGN SOLUTIONS

The muscle action involved in pushing the levers at chest level is more direct than rotating the arms at hip level, and therefore requires much less strength. The energy produced propels the chair 10 percent faster than conventionally operated wheelchairs. The two levers are operated independently to facilitate turning, while eliminating the problem of contact with dirty tires.

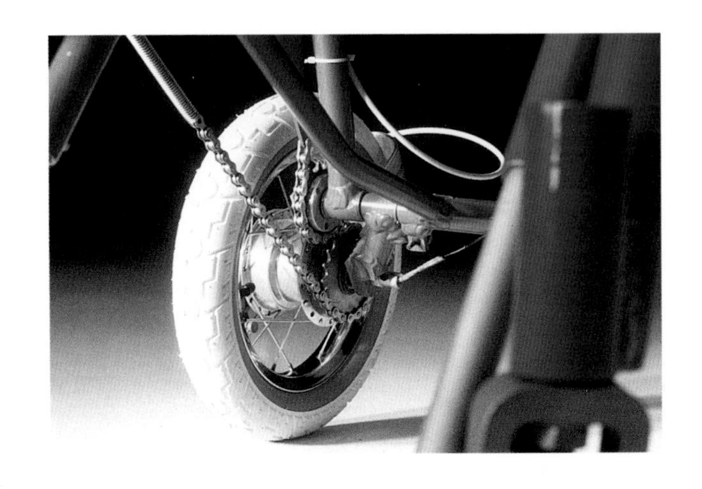

DESIGNER
Douglas D. Clarkson
DESIGN SCHOOL
Art Center College of Design

"...This is innovative, paradigm-busting, problem/solution thinking coupled with a technical fluency rare enough in the professional world and altogether exceptional in a student project."

JURY CHAIR ARNOLD WASSERMAN, IDSA

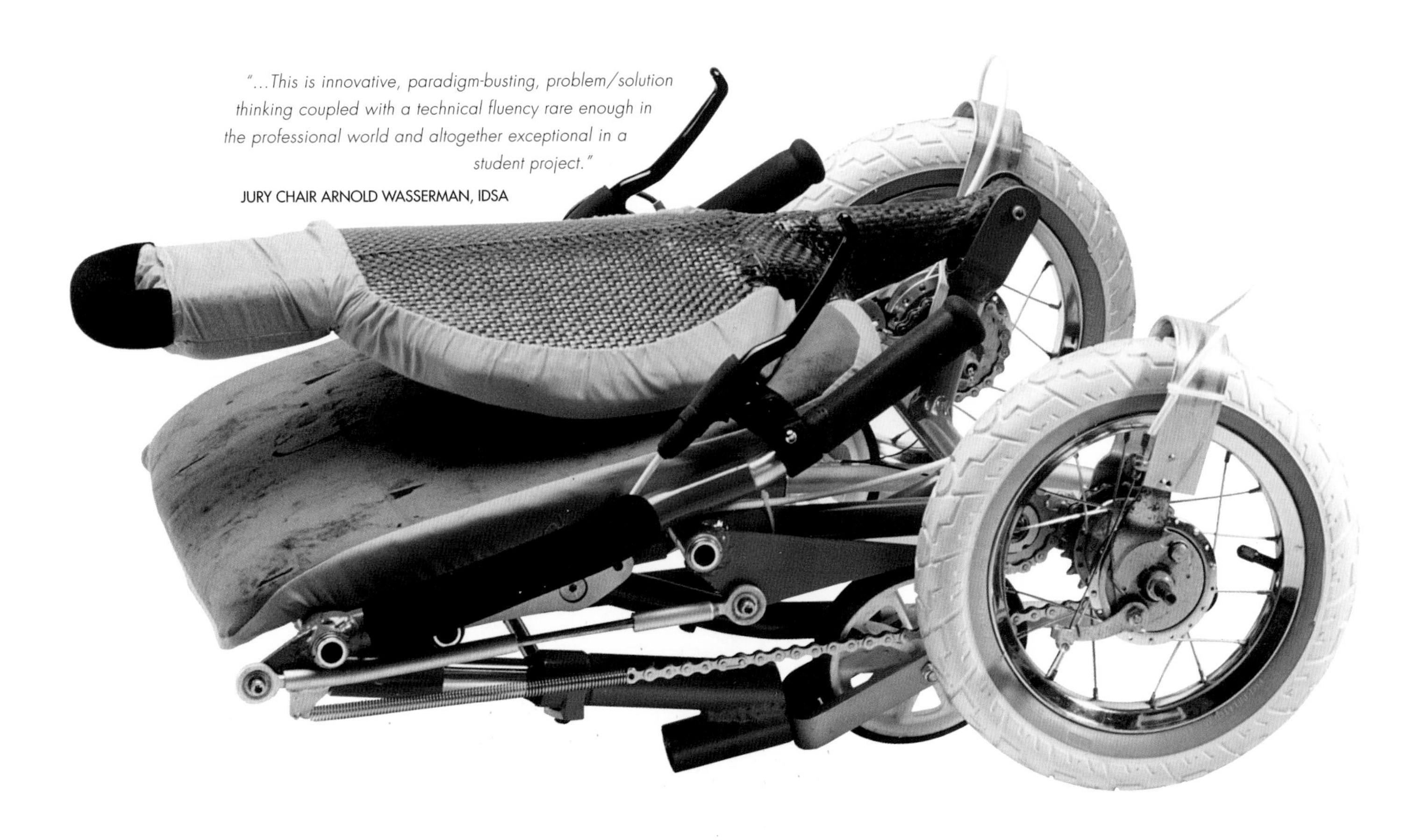

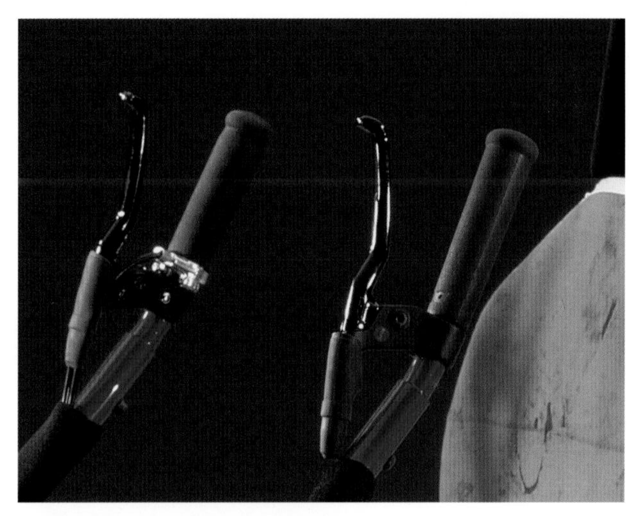

RESULT

The New Move is propelled by a lever (or rowing) action, which is 100 percent efficient versus the 40 percent efficiency you get pushing a wheel.

Seville STS

DESIGNERS
Charles M. Jordan, John R. Schinella,
Dennis Little, Marv Fisher of
General Motors Design

CLIENT
Cadillac Motor Division,
General Motors Corp.

"The Seville is a successful execution of a high-risk strategy: to develop a new design statement for an established luxury vehicle. They achieved the objective without resorting to a look-alike, Euro-luxury solution. It is a clean and restrained design with refined detailing, yet it has a definite, but understated, presence on the road."

JUROR FRITZ MAYHEW, IDSA

DESIGN OBJECTIVE

To create a Cadillac with dramatic new looks and international appeal combined with greater interior spaciousness and comfort.

DESIGN SOLUTIONS

To achieve this level of quality, Cadillac employed the latest technology in its manufacture. For example, special cameras ensure an accurate and sound build by performing a "visual inspection," checking the fit and alignment of body panels which are precision welded by lasers and robots.

RESULT

A well-balanced, internationally appealing sedan, Cadillac's new Seville STS is part of a long-range investment in quality rather than a short-term, high-volume sales story; nonetheless, it has exceeded all sales and marketing projections, with retail sales up 89.5 percent in the US in the first six months. In fact, foreign sales have tripled.

OTHER AWARDS

–*Motor Trend*, Car of the Year
–*Automobile*, Automobile of the Year
–*Car and Driver*, Ten Best List

The Brick

DESIGNERS
Scott H. Wakefield,
John E. Thrailkill of
Bleck Design Group

CLIENT
Ergo Computing Inc.

DESIGN OBJECTIVE

To develop a transportable computer with the power and flexibility of a desktop PC and the mobility of a laptop that would allow the user to transport only the central processing unit while moving between keyboards and displays at different locations.

DESIGN SOLUTIONS

A spattered finish was chosen because of its aesthetic interest and ability to hide any abuse a portable product might get. By combining liquid heat sink technology with the extruding body acting as a cold plate, a slower fan was required. The combination of increased convection and forced air results in extremely quiet operation.

RESULT

The main case is an extrusion, allowing Ergo to build the computer in two lengths without modifying the other case parts. Made up of several components, The Brick can be configured by the user for maximum comfort–an advantage unique in the portable PC market.

"The rugged industrial strength personality of The Brick is completed by the molded bumper caps at the corners of the CPU and the flat screen. These beautifully integrate the ensemble."

JUROR ARNOLD WASSERMAN, IDSA

Good Grips Kitchen Utensils

"The Good Grips product line is a brilliant example of universal design–it satisfies the needs of people with reduced grip strength, while appealing just as strongly to other consumers...."

JUROR LIZ POWELL, IDSA

"We were eager to demonstrate that designing for the elderly was not an excuse to make frumpy prosthetic devices, but rather an opportunity to make things better for everyone."

DAVIN STOWELL, PRESIDENT OF SMART DESIGN INC.

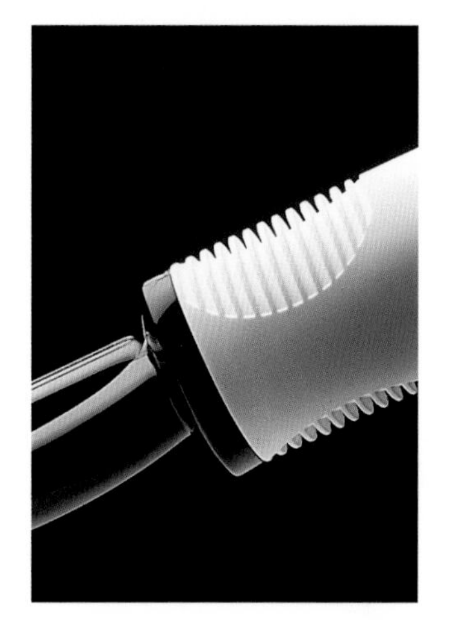

DESIGN OBJECTIVE

To create sensible, high-performance, easy-to-use kitchen gadgets.

DESIGN SOLUTIONS

Common user problems were overcome by forming the tools to spread out stress, involve the whole hand and muscles, and allow for a wide variety of grip positions. The larger diameter increases leverage, while the large oval section keeps the tool from rotating in the hand. Big, easy-to-see, color-coded denominations mark the measuring devices.

RESULT

The central facet of Good Grips is the big "rubber" handles with fins which provide a firm grip, and the wide, flat handles and knobs on the squeeze tools, which spread pressure across wide areas of the hand.

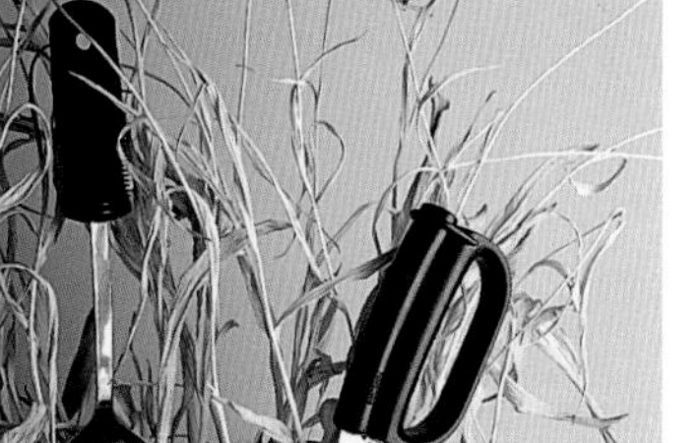

OTHER AWARDS

–*I.D.* Annual Design Review
–*International Design Yearbook*
–*Metropolitan Home*, 100 Best Selection
–London Design Museum's '92 Review

DESIGNERS

Davin Stowell, Dan Formosa, Tucker Viemeister, Michael Callahan of Smart Design Inc.

CLIENT

Oxo International

PEAK 10 CAMP STOVE

DESIGNER
Brett Ritter
FACULTY ADVISORS
Michael Kammermeyer, Herb Tyrnauer
DESIGN SCHOOL
California State University–Long Beach

"This is an uncommonly thoughtful solution to a relatively common project. The designer clarified the contributions design could make within the existing camp stove paradigm and executed them with responsibly innovative design solutions...."

JUROR PETER W. BRESSLER, FIDSA

DESIGN OBJECTIVE

To develop a camp stove with a minimum 13-inch spacing between the two burners, a minimum fuel capacity of six pints, electronic arc ignition, a more simplified operational format, and an overall reduction of parts.

DESIGN SOLUTIONS

A clam-shell or folding-halves design proved the best way to increase burner spacing, while requiring less storage space, and twin tanks achieved greater fuel efficiency. A simple piezo crystal arc ignition allows for safer lighting. For ease of use, the modular components are dishwasher safe.

RESULT

The Peak 10 Camp Stove comes in at $2 below the cost of manufacturing for the old Coleman two-burner unit. It has a strong visual identity and an intuitive operating procedure.

Macintosh Color Classic

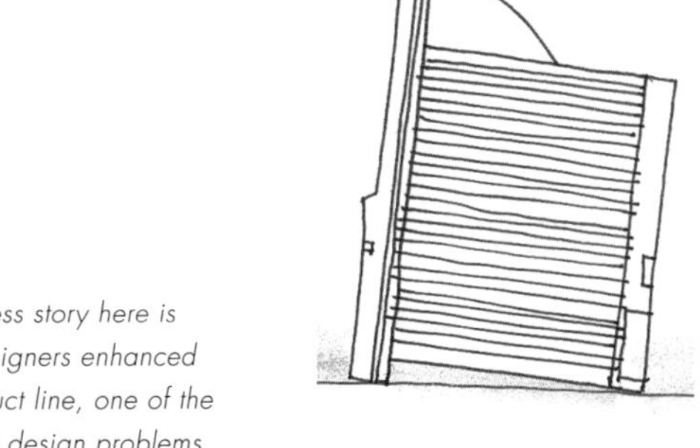

"The major success story here is how well the designers enhanced an existing product line, one of the most challenging design problems facing a maturing business....with the right mix of user-sensitive features, clever design details and a fresh aesthetic...."

JUROR LOU LENZI, IDSA

DESIGNERS
Larry Barbera, Daniele DeLuliis,
Robert Brunner of Apple Computer, Inc.

CLIENT
Apple Computer, Inc.

DESIGN OBJECTIVE

To rewrite the Apple design language, since the Snow White language that had been established in 1984 was no longer unique, having been replicated by competitors.

DESIGN SOLUTIONS

Mechanical requirements of attaching the bezel to the bucket to pass drop tests fostered the rear "bubble." This made it inboard from the sides to allow straight-through screw clearance and eliminate visible tunnels. An air space beneath the front bezel gave the product an enhanced sense of lightness and vitality to offset the increased height from the 10-inch Trinitron color tube.

RESULT

The design integrates the handle over the Classic's center of gravity for improved balance when lifting. The product uses a very low emission CRT for customer safety.

OTHER AWARDS

–Industrie Forum Design Hannover, 1993 iF Award
–*I.D.*, 1993 Award

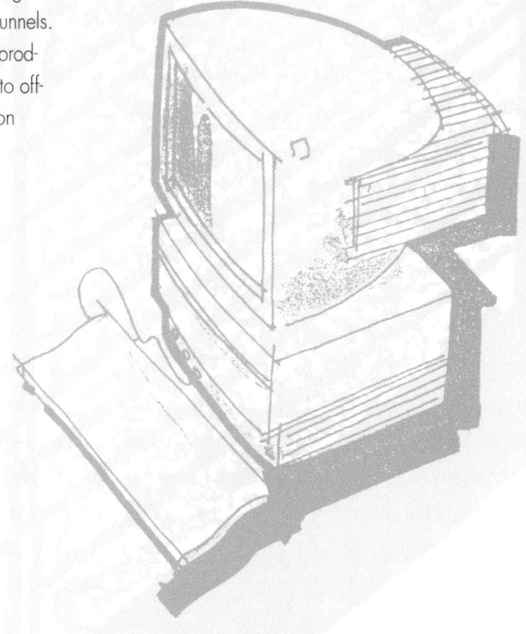

JuiceMate

DESIGNER
Bart André
ADVISOR
Michael Kammermeyer
DESIGN SCHOOL
California State University–Long Beach

DESIGN OBJECTIVE

To design an extractor that was more efficient, compact, and accurately proportioned than those currently on the market.

DESIGN SOLUTIONS

The two-part filter/blade unit is easy to clean and protects the user from injury. Moreover, the design incorporates both a reamer and an extractor, to allow the JuiceMate to work for both vegetables and citrus fruits. It uses older, more efficient technology to extract juice than its would-be competitors, but it uses that technology in a new way, with an injection-molded blade/filter to reduce costs.

RESULT

All of the changes were made while still keeping the product in the $40–$90 price range. Added costs, from parts like the citrus reamer and motor housing, were defrayed with other aspects of the design, like the elimination of the pulp container and the added value of the citrus reaming feature.

"It is wonderful and rare when it becomes difficult to tell student from professional work. This sumptuous little juice maker has more eye appeal than most such juicers now in production."

JUROR CHARLES BURNETTE, Ph.D.

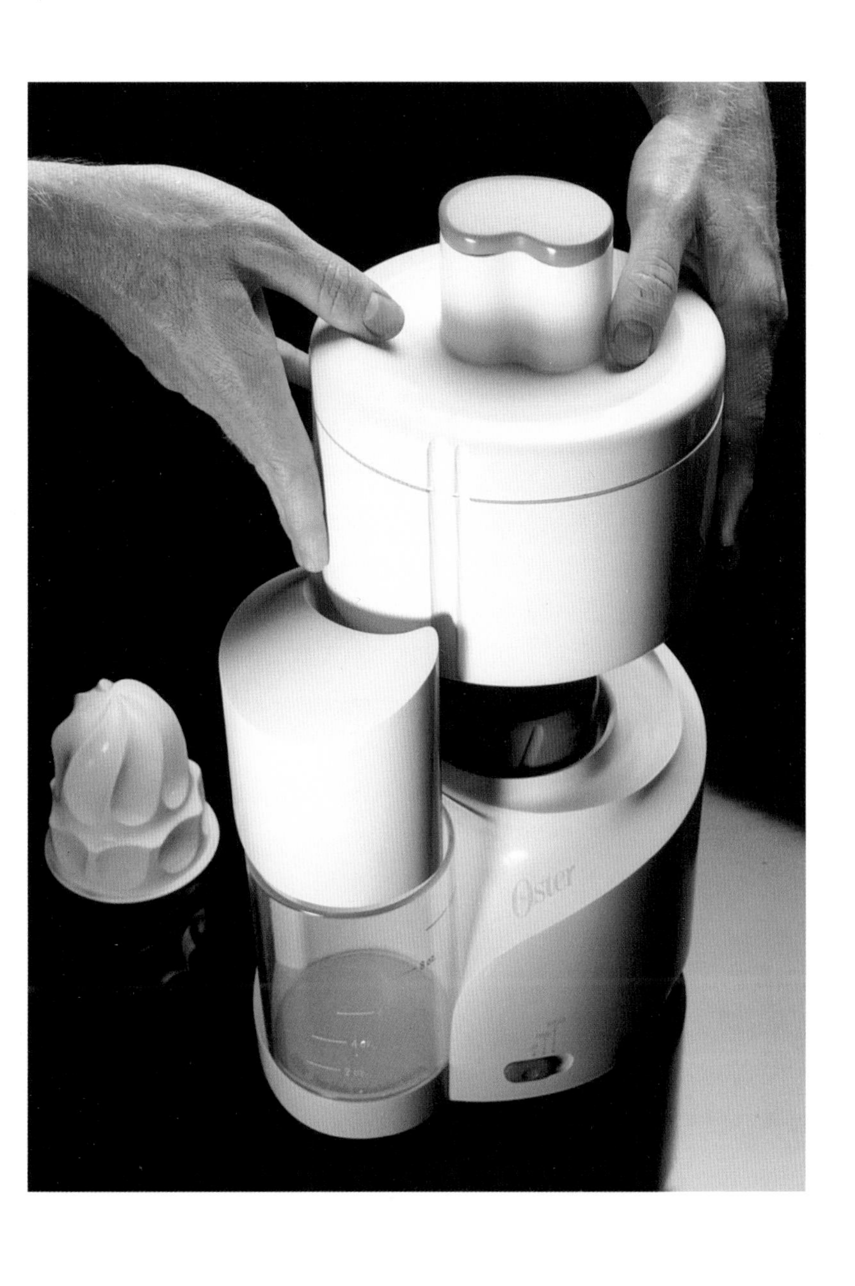
Oster

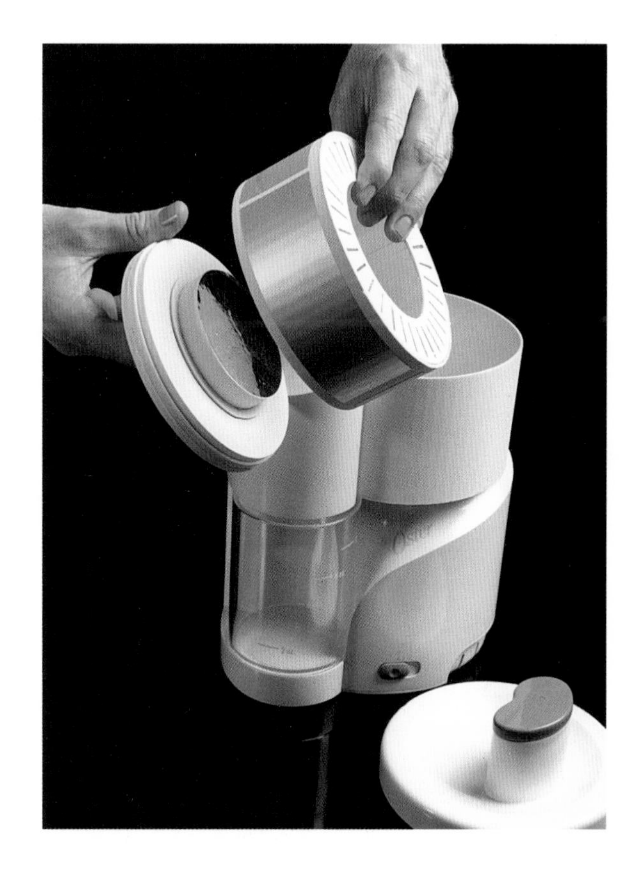

LH Midsize Sedans—

Dodge Intrepid, Eagle Vision and Chrysler Concorde

DESIGNERS
Thomas C. Gale, Trevor M. Creed, John E. Herlitz, K. Neil Walling and The Chrysler Design Staff of Chrysler Corp.

CLIENT
Chrysler Corp.

"Smooth, well-thought-out exterior forms and surfaces, combined with a larger interior space offering better ergonomics: The LH Program cars reset the bar much higher as a target for competitors in their market."
JUROR CHARLES ALLEN, IDSA

"...Where else can you get a full-size car (120 cubic feet interior volume with an overall length that of the standard-wheelbase Mercedes S-Class) with a mid-size price, from base Intrepid at $15,000 to loaded Vision at about $23,000?"
AUTOWEEK, MAY 4, 1992

DESIGN OBJECTIVE
To differentiate GM-10 midsize four doors from their competition in both functional and aesthetic appeal.

DESIGN SOLUTIONS
The LH Program's cab-forward architecture increases interior space, has larger doors for ease of entry and exit and a large glass area for 360-degree visibility. The windshield was moved forward and down from the normal sedan position, reducing wind noise and improving fuel economy.

RESULT
The success of the cab-forward LH family through June of 1993 has produced a 33 percent passenger car sales increase for Chrysler over the same period for 1992. By 1995, all Chrysler's domestically produced four-door sedans will feature the cab-forward vehicle architecture pioneered in the LH cars.

Silhouette Window Shadings

DESIGNERS
Wendell Colson, Paul Swiszcz, Joe Kovach,
Terry Akins of Hunter Douglas Window Fashions Division

CLIENT
Hunter Douglas

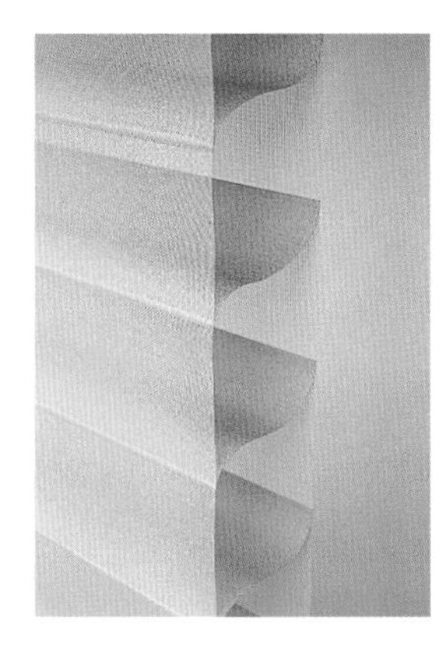

DESIGN OBJECTIVE

To create a window treatment with the light control capability of a venetian blind combined with the softness of a curtain and the easy operation of a shade.

DESIGN SOLUTIONS

A continuous cord loop pulley system raises and lowers the shading. With a lifetime warranty, the system's slats are vanes that flex from a flat plane to a gently s-shaped curve as they are tilted. An integrated headrail, valance, bracketing system controls the shade roller inside the sleepily curved valance. Instead of the traditional end-mount brackets, the design team used brackets associated with pleated shades for a more elegant look.

RESULT

The shading blocks 98 percent of UV rays when the vanes are closed and 68 percent when open. First year sales have been excellent, approaching $100 million retail in the US; it has achieved over 50 percent growth annually since then.

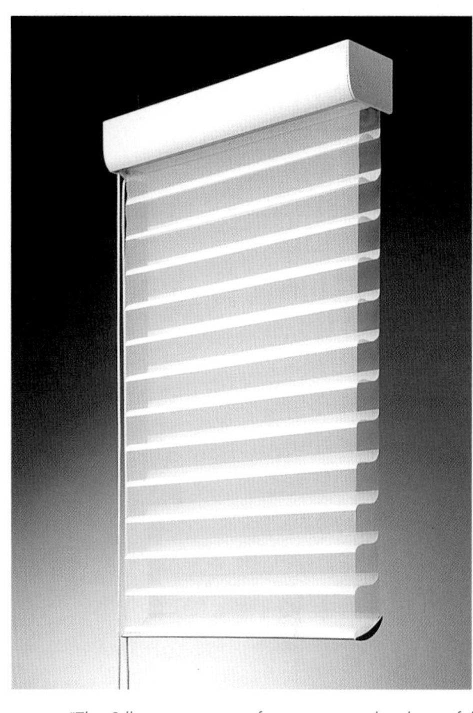

"The Silhouette is a terrific concept with a beautiful application of materials. Its rational design and combination of comfort with elegance earned a unanimous thumbs up from the jury."

JUROR VINCE FOOTE, FIDSA

Sensor for Women

DESIGNER
Jill Shurtleff of
The Gillette Co.

DESIGN OBJECTIVE

To create a break-through handle shape that addresses the specific needs of female shaving while enhancing the Sensor blade technology.

DESIGN SOLUTIONS

The Sensor for Women's patented, ergonomically designed handle breaks the tradition of women's razors as pastel-colored versions of male products. It provides better feedback, control and maneuverability, eliminating the handle roll common with traditional razors. The clear grip, made of impact-modified acrylic, enhances the product's water imagery and gives the razor its correct weight and balance.

RESULT

Within its first three months, one out of every two refillable razors sold (including men's) was a Sensor for Women. It has been the best-selling razor since it attained full distribution in August '92.

"...The design's blend of human factors, proper materials selection, right manufacturing processes and appealing aesthetics make it a major contributor to both Gillette and the quality of life of the women who use it...."

JUROR LOU LENZI, IDSA

ECO35

"...Designers should build in solutions to social problems. Obviously a beautiful form, this cute little camera goes beyond fulfilling the requirements–the designer has redefined the disposable paradigm...."

JUROR TUCKER VIEMEISTER, IDSA

DESIGN OBJECTIVE

To create a simple 35mm camera with optical quality similar to the single-use products, preloaded with film and sold at a comparable price to the disposable cameras.

DESIGN SOLUTIONS

The oversized film advance thumbwheel, film rewind crank, large shutter release button and neck strap were designed to make operation obvious. A removable sticker across the film door latch reminds the user that the camera is preloaded, preventing accidental exposure.

RESULT

The ECO35 is a disposable camera with a difference. The chipboard packaging is of recycled material, so that the product's overall environmental impact is minimized.

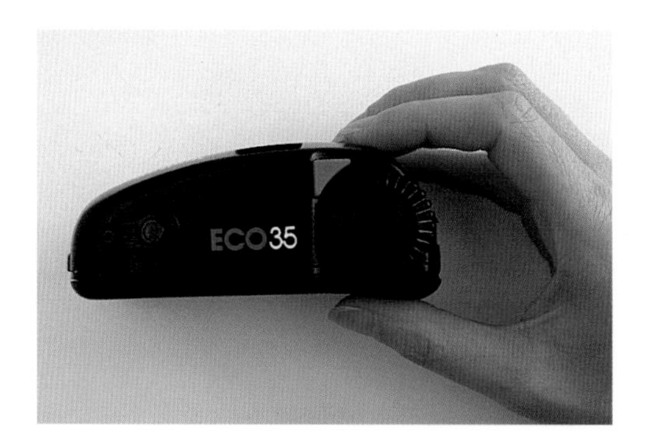

"For $19.95, less than many 35mm disposable cameras, it includes a 24-exposure roll of Kodacolor Gold 200 film, which comes preloaded with the camera. Later you can use any print film, from ISO 100 to 400."

JUDITH ARANGO,
MERCURY NEWS

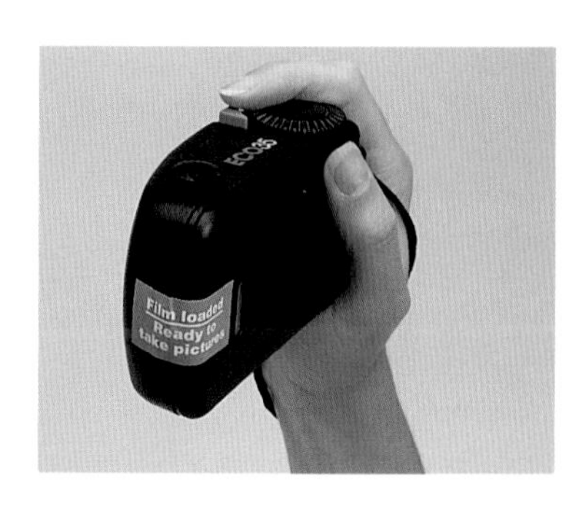

DESIGNERS
Steven Shull of Vivitar;
Alan Stone, James Flynn of
Alan Stone Creative Services

CLIENT
Vivitar Corp.

Telesketch

This designer's brainstorming tool lets the caller and receiver create sketches and ideas simultaneously, facilitating the conceptual stage of design in today's international corporation. The button and receiver docking area radiate from the handset dial.

DESIGNER
Tark Abed

DESIGN SCHOOL
California State University–Long Beach

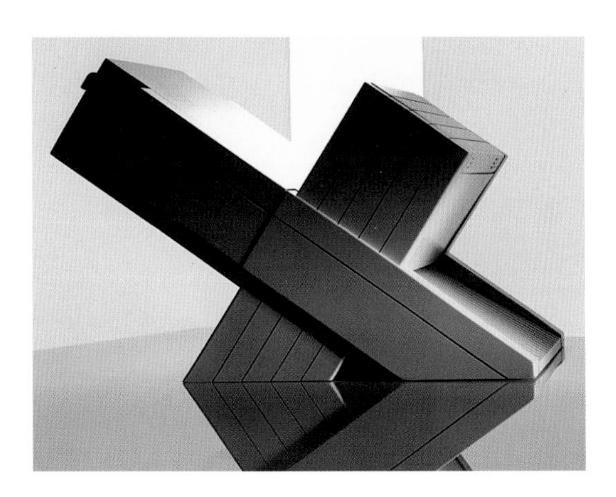

AddressWriter

DESIGNER
John Lai of Matrix Product Design

CLIENT
CoStar Corp.

The AddressWriter is the first desktop printer specifically designed to print envelopes and postcards. The *X* shape elegantly presents the envelope directly to the user and provides easy access to the envelope input in the rear. Intuitive to use, the design takes minimal desk space and features a snap-together casing that requires only four screws for the entire assembly.

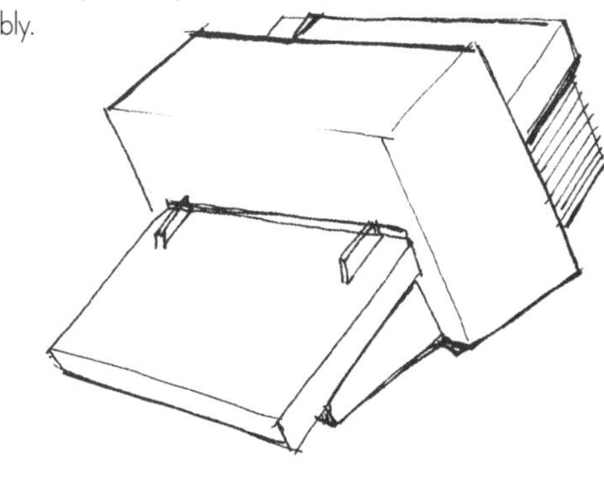

MOTO 1 Motorcycle

DESIGNER
Robert Steinbugler of
Space, Form and Structure

CLIENT
MOTOcycles

The MOTO 1 is a high-end motorcycle that bridges the age-differentiated motorcycle market. Rather than an assemblage of parts, the MOTO 1 reads as a unified vehicle because of its innovative bodywork. The design allows the wheel to travel up and down within the body but articulates the form along the axis of the steering head to accommodate left-to-right movement— an innovation unique to the industry.

Contour

DESIGNERS
Jack Telnack, Chuck Haddad,
Tom Scott, Pat Schiavone,
Taru Lahti, Cecile Giroux,
Ken Grant of Ford Motor Co.

CLIENT
Ford Motor Co.

The Contour's design explores the leading edge in automotive technology in a new wave, global luxury sedan. Its bonded aluminum space frame demonstrates new construction techniques that allow lightweight, high-strength, low-investment body structures. With maximum space inside, the design offers more easily accessible engine components. Both the space frame and plastic body panels can be recycled.

Stingray Trackball

DESIGNER
Paul Bradley of Matrix Product Design

CLIENT
CoStar Corp.

This trackball provides a much more comfortable input device for the Macintosh. The design lowers the entire device to a comfortable height by using a smaller ball and contouring the buttons down to their lowest possible height. By giving the buttons 70 percent of the Stingray's surface, the design accommodates a wide range of hand sizes and use styles. Forty percent cheaper to manufacture, it is also much smaller and lighter in weight than other trackballs.

sHand-Tite™ Keyless Chuck

DESIGNER
Robert Huff of The Jacobs Chuck Manufacturing Co.

CLIENT
The Jacobs Chuck Manufacturing Co.

This chuck redefines the performance and ease of operation of portable power drills. Eliminating the chuck key usually needed to secure the drill bit by hand, a time-consuming and cumbersome operation, the Hand-Tite provides fast and easy bit changeovers and improved drilling accuracy. First year sales exceeded 100,000 units and 1991 unit sales were projected at over 750,000. Moreover, the ergonomics makes it easier for disabled and elderly individuals to use.

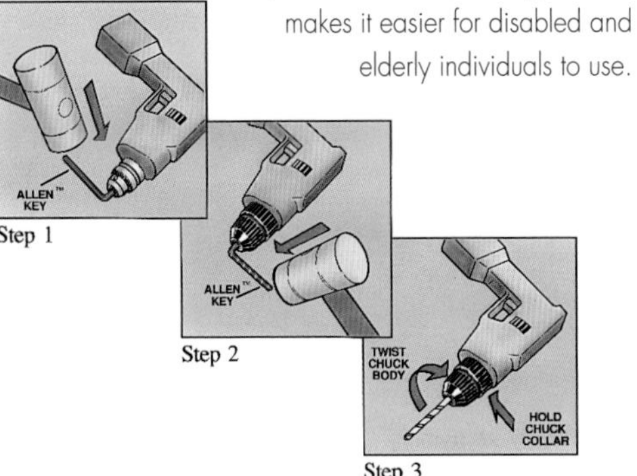

Spoons to Grow With

DESIGNER
Adam Straus of Adam Straus Co.

CLIENT
Adam Straus Co.

Design research shows that the traditional narrow spoon handle is inappropriate for the overhand grip of babies and small children. Instead, these spoons provide a broad surface with an undercut or relief in their underside to enable kids to get a firm grasp and control of the spoon. Animals, a recurring theme with babies, are shown in action, providing a fun, playful motif that children love to use.

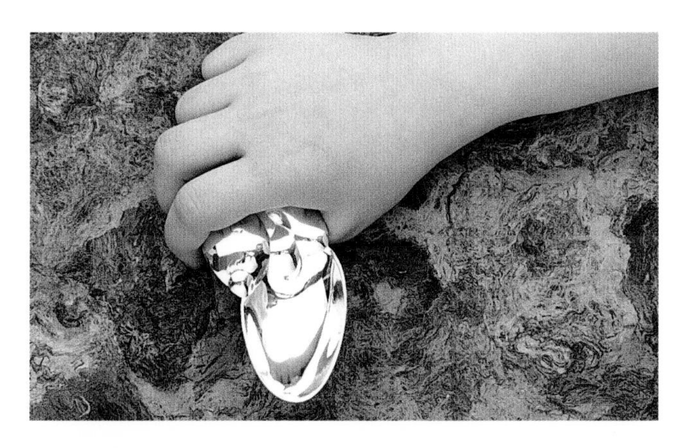

Squejé® Aqua Tooth Polisher®

DESIGNERS
Peter H. Muller of Interform;
George G. Michaels, DDS of Squejé Corp.

CLIENT
Squejé Corp.

Designed to accommodate the needs of users with a full range of abilities, the Squejé® polishes teeth and massages gums. The handle's design integrates a sculptured *S* curve with thumb/finger rests and a generous hand grip/palm rest to support the best positioning of the Squejé® head against the teeth and gums. With functional prototypes lasting for almost three years, the product produces less waste.

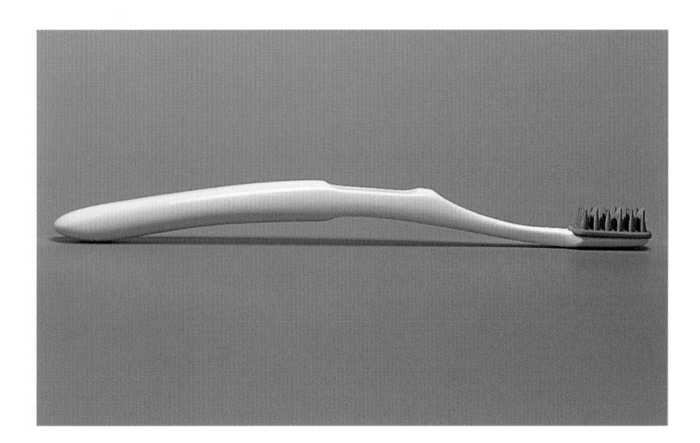

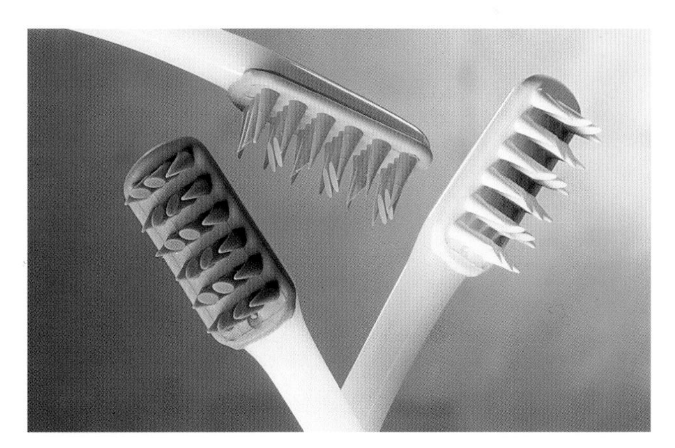

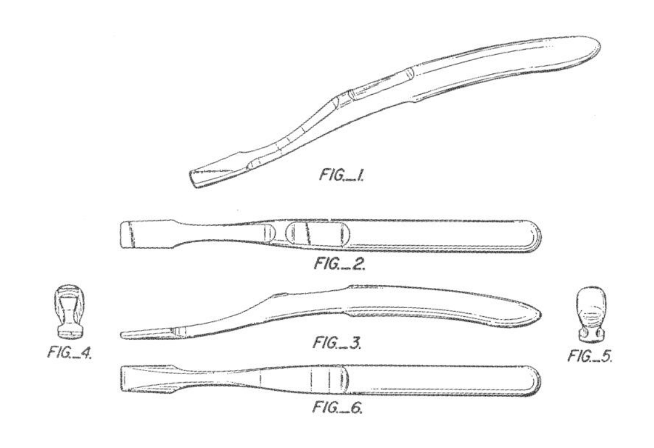

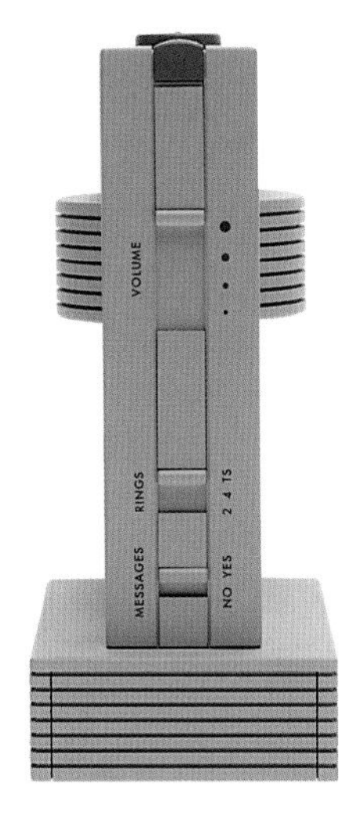

Digital Answering System 1337

DESIGNERS
Dan Harden, Paul Braund of frogdesign inc.

CLIENT
AT&T

This product stores messages on a chip, making playback, repeat, forward and delete instantaneous. The design expresses the new all-digital technology while providing intuitive use. With no moving parts, it uses less energy to operate and, because of its size—a footprint ¼ the size of competing products—it requires less plastic in fabrication.

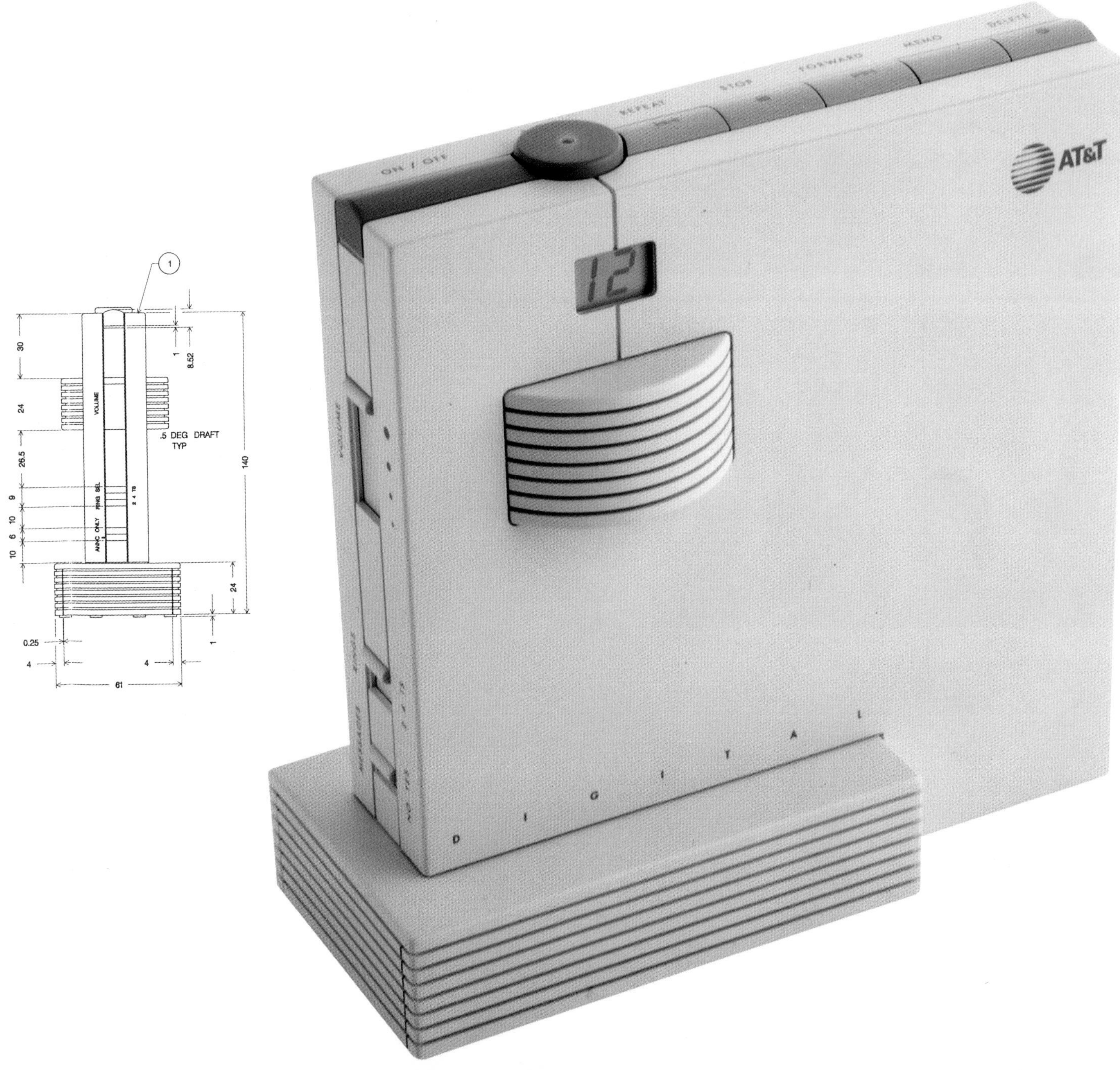

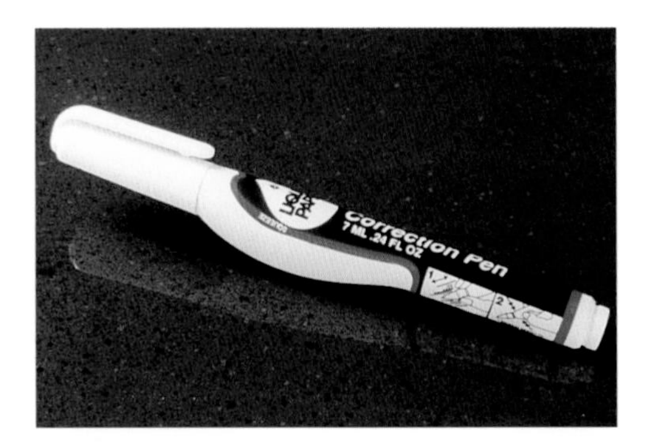

Ergonomic Correction Pen

DESIGNERS
Bryce Rutter of Herbst LaZar Bell Inc.;
Norm Poisson, Ron Draper of
The Gillette Company,
Stationery Products Group

CLIENT
The Gillette Company,
Stationery Products Group

This pen contains liquid paper correction fluid, dispensing the fluid when the pen tip is pressed against paper. By squeezing the belly of the pen, the user controls the amount of fluid dispensed. The design derives from careful analysis of hand anatomy and provides optimum control and coordination over the pen by fingers. The design is both cost-effective and easy to produce in high volumes.

GRiDPAD

DESIGNERS
Christopher Loew, Robin Chu of ID TWO;
Jack Daly of GRiD Systems Corp.

CLIENT
GRiD Systems Corp.

The form of this handwriting recognition computer can be comfortably carried and used like a clip-board. The cord spool acts as a waistline for the device. Below it, the main body is soft and it can be cradled in the arm.

Flexor® Gloves

DESIGNER
Dixie Rinehart of
Rinehart Glove Ltd.

CLIENTS
Swany USA; The US Army;
American Honda Co.

This glove incorporates a patented pattern for the finger construction. With 50 percent fewer seams than conventional gloves, that pattern permits the hand to flex like the natural human hand, thereby providing improved comfort, warmth and durability. It takes 50 percent less energy to flex the fingers of this glove.

Attaché Portable Computer

DESIGNERS
Andrew Serbinski, Daniele Delullis,
Peter Von Maydell of Serbinski Machineart;
Dana King of Master Model Makers

CLIENT
Fujitsu Ltd.

This concept, a portable computer compatible with Unix-based PCs, is placed on a surface and the leather cover flips open. The keyboard swings down and the display swings out to a suitable angle. With an exterior of leather, the design seeks to define the characteristics of quality personal accessories and to translate these attributes visually as an expression of technical sophistication.

Key Chain Security Remote Transmitter

DESIGNERS
Mark Dziersk, David Harris, Tim Repp of
Group Four Design

CLIENT
Audiovox Corp.

Used to activate and disarm the Audiovox automated car security system, this remote control's shape makes it easier to recognize by feel in a purse or pocket. The code buttons are different sizes and colors so that they can be distinguished visually and by feel in the dark. Immediate orders were up 20 percent over orders a year ago. Moreover, the design of the units also reduced materials to a minimum, resulting in a more profitable product for Audiovox.

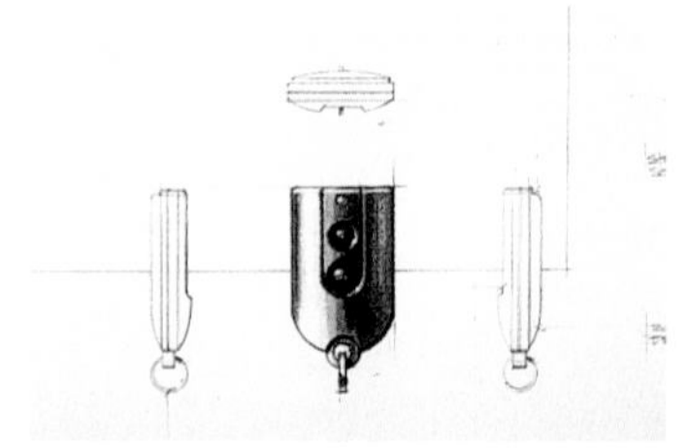

1-2-3 Bike

DESIGNERS
Richard Daley, Fred Rieber, Joe Sejnowski, Dave Wohl of Preschool R&D, Playskool Division of Hasbro

CLIENT
Playskool Division of Hasbro

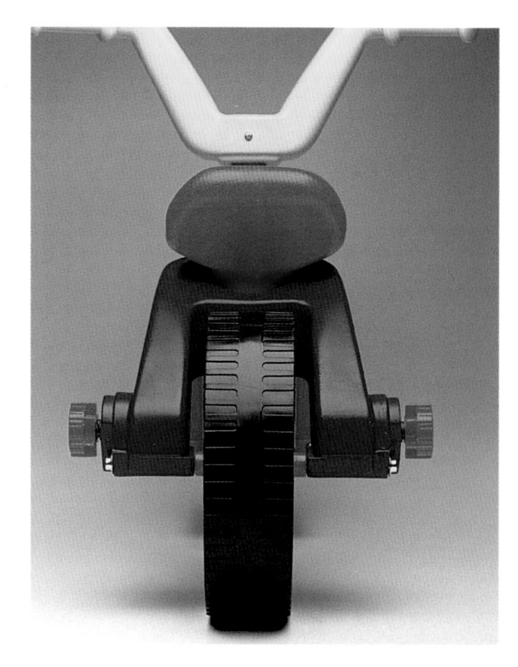

The 1-2-3 Bike provides a unique three-stage training system for preschool bike riders. Its training wheels and seat height adjust easily without tools. As the child's balance develops, the training wheels can be adjusted inwardly, without raising them from the ground like traditional training wheels. The result is a more stable training bike that provides a gentle transition to two-wheeled riding. Its design made the 1-2-3 Bike the best selling ride-on toy within five months.

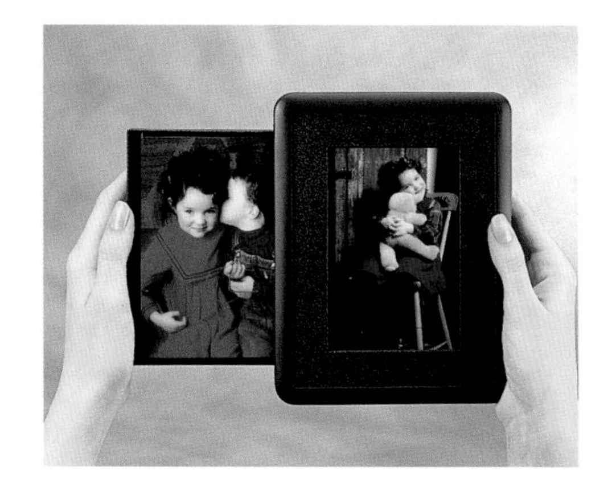

Showbox Photo Viewer

DESIGNERS
Hanspeter Leins of frogdesign inc.;
Peter Ackeret of Showbox Systems AG;
Diego Bally of D&A Design and Advertising AG;
Wendy Segnalla of Design Continuum Inc.

CLIENT
Burnes of Boston

The Showbox holds up to 40 photos which easily load into a concealed drawer. When closed, the top photo appears in the window. To view the rest of the photos, the user simply opens and closes the drawer and Showbox automatically advances to the next photo.

"Goldilocks and the Three Bears"

DESIGNERS
Robin Chu, Daniele DeIullis of ID TWO;
Mike Musel of Precision Prototypes and Models

CLIENT
Apple Computer, Inc.

In response to a commission to explore appearance and configurations for input devices that will be relevant to future generations of Apple products, ID TWO designed a keyboard that will store in a minimum volume for portability and that opens up to the butterfly position to accommodate user comfort. The puck shape is smaller and easier to point than a conventional mouse.

Kidz Mouse

DESIGNERS
Dan Harden, Tino Melzer of frogdesign inc.

CLIENT
Logitech, Inc.

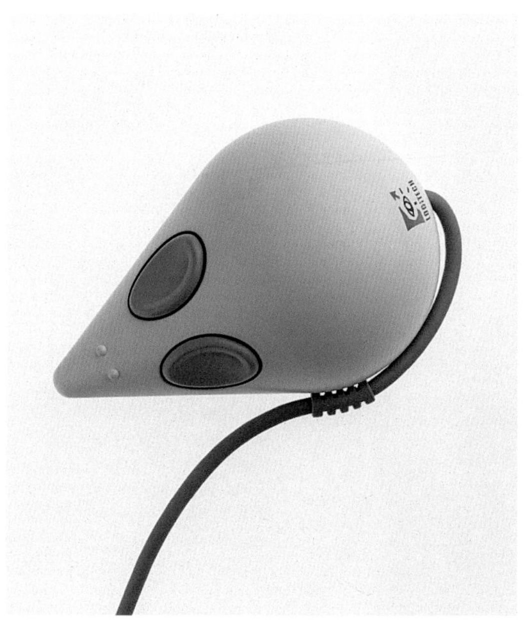

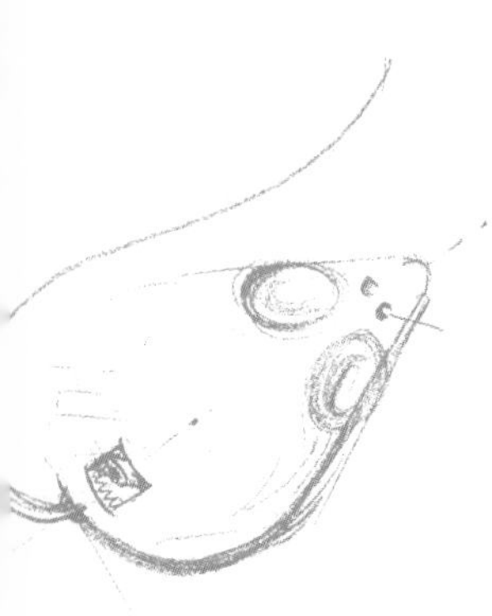

Designed specifically for kids ages 4 to 12, the Kidz Mouse looks like a real mouse, complete with eyes, ears and tail. Since children do not have precise dexterity in their fingers, this mouse has two large cupped ear buttons and a domed body for the hand to rest on. Made smaller to fit smaller hands, the mouse's pointed nose helps children orient and direct the mouse.

Momenta Pentop Computer

DESIGNERS
Paul Braund of Braund Creative Design, Inc.

CLIENT
Momenta Corp., Inc.

The Momenta combines—in a space smaller than a three-ring binder—all the features of a laptop, pen-based notebook, PC and fax machine. An innovative soft cover casing over an impact-resistant hardcase replaces the conventional hard plastic housing. The system can be used a variety of ways—sitting up, lying down, or folded out.

’vik-ter Stacking Chair

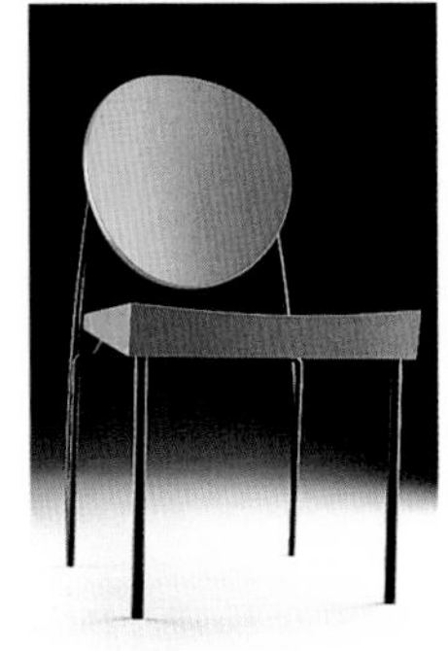

DESIGNERS
Dakota Jackson, M. C. Graves of Dakota Jackson Inc.

CLIENT
Dakota Jackson Inc.

For this eye-catching chair with its uniquely sculptural form, the designers selected cherry—a domestic wood that's not endangered—and developed color stains inspired by Matisse's paintings of Morocco (1912-1913). The gently curving steel frame allows the seatback to move forward and backward and the conical rubber springs joining the backpan to the frame let the seatback follow the body in a range of positions.

Image RV Awning System

DESIGNERS
LeRoy J. LaCelle, Gregory Marting, Kurt Solland, Tony Grasso of Designhaus, Inc.

CLIENT
Carter Shades

Based on analysis of actual awning operation and a study of injury reports, the design offers older adults a user-friendly, self-explanatory system. Unlike its competitors, it is designed as a wholly integrated product with soft, rounded extrusions blending into the aerodynamic protective roller caps.

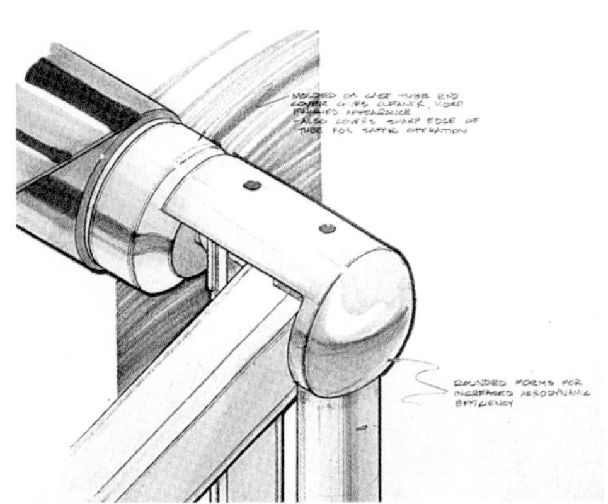

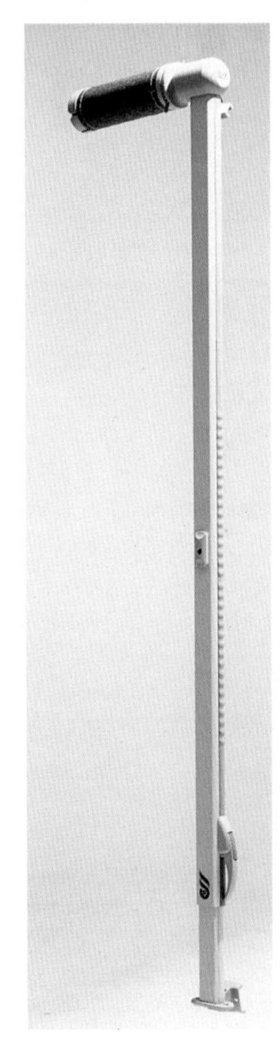

Deni Freshlock-Vacuum Sealer

DESIGNERS
Gregg Davis, John Koenig,
Tim Friar of Design Central

CLIENT
Deni/Keystone

This product extracts air from plastic bags that have food in them to increase their storage life. The pressing movement activates the vacuum fan and the heat sealing wires. While the product's form prevents liquids from spilling into the unit, a further safeguard is provided by the three wires—one to seal the bag in use, one to seal the next bag and the third to cut the two bags apart.

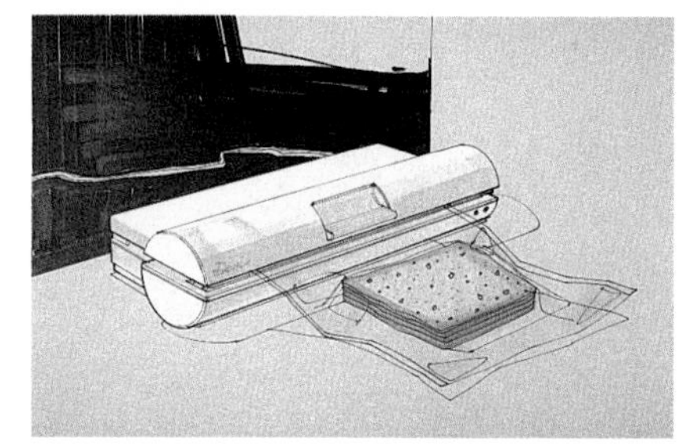

Plastic Combination Lock

DESIGNERS
Martin Smith, Leonard Porche

The first plastic combination lock design, this material will resist all but the most determined efforts to tamper, like metal; however, unlike metal, the plastic needs no lubrication and will not corrode, giving the product a far longer life span. A watch battery supplies the power to backlight the numbers, improving access at night.

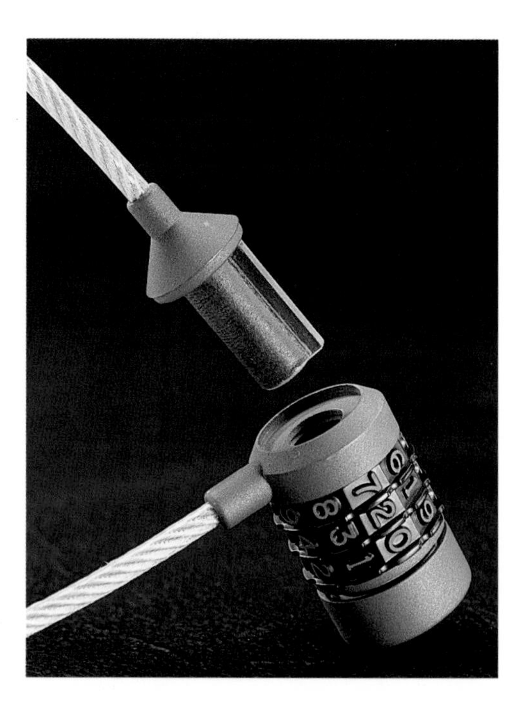

Spacecab Medicine Cabinet

DESIGNER
Mark Andersen of S.G. Hauser Associates
CLIENT
Zaca Investments Inc.

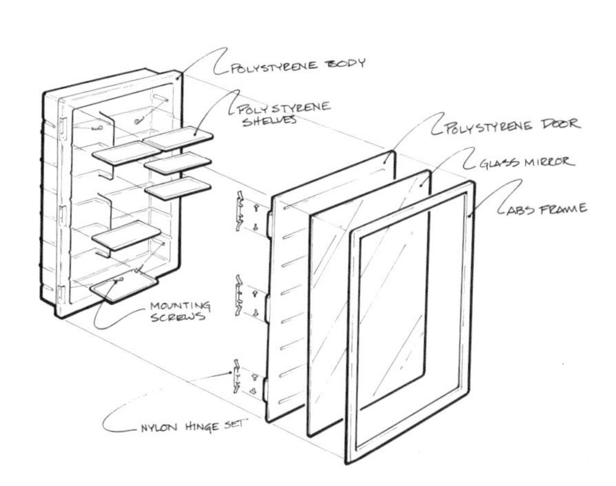

Made of injection-molded thermoplastics, the Spacecab requires no painting and degreasing—minimizing hazardous materials in manufacture. This cabinet requires no hardware or tools to assemble, and it won't rust like traditional cabinets.

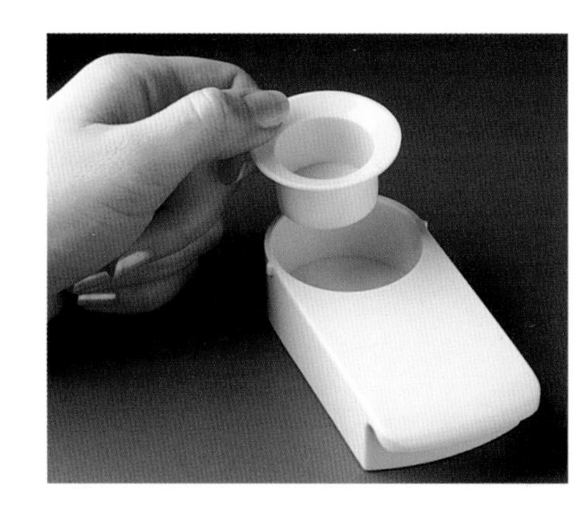

Portions™

DESIGNERS
Wayne Husted of Wayne Husted Industrial Design
CLIENT
TCB Inc. Concepts Plus Div.

Portions™ stores and dispenses dry foods. After the jar is filled, it is stood on its head. Removing the built-in scoop activates a cut-off blade, coupled with a precise system of springs, catches and releases, allowing the contents to be removed easily in fresh and accurate quantities. First shown at the housewares fair in Frankfurt, Germany, distributors from 14 countries have made initial commitments and the US client placed an initial production order for 65,000 units.

Seatcase Compact Wheelchair

DESIGNERS
Joseph M. Jarke, Ole I. Thorsen of Jarke-Thorsen Products, Inc.; The Design Team of Beach Mold & Tool; The Design Team of GE Plastics

CLIENT
Jarke-Thorsen Products, Inc.

Lightweight, durable, highly maneuverable and easy to manufacture, this wheelchair can be carried with one hand, in the lap or on the back of a standard wheelchair. Made with a unique folding mechanism that makes it compact and strong, the Seatcase fits through narrow doors and pathways. It opens up in seconds and can even be used in a shower.

Direct Manual Braille Slate

DESIGNERS
Lawrence S. Hawk, Joe H. Turner, Joyce Maienschein, Regie Linsey, James R. Palmer of Martin Marietta Energy Systems Inc.

CLIENT
Matrix Tool & Mold Co.

This dramatically innovative product is used to code braille directly, left to right, on the front side of the page, opposite to the way other braille writing products are used. The hollow stylus forms the dots on the front side of the page, drawing the paper into a cylinder with a spherical top about eight times as high as the paper thickness. Visually impaired persons have been able to transcribe 75 to 100 words per minute with this unit.

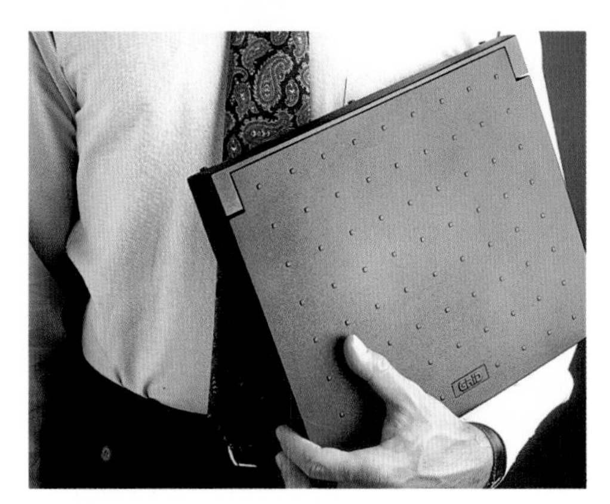

GRiD 1810

DESIGNER
Neil Taylor of ID TWO

CLIENT
GRiD Systems Corp.

This notebook computer marks a transition from the aesthetic of the company's first product to a more consumer-driven appearance. The design makes the removable hard drive feature—the key advantage of the product—easy for the user to take full advantage of, with an elegant and easy method of operation.

Microphone and RCA Adapter

This microphone gives users the option to input voice to a Macintosh. The design works on the desk, in the hand, clipped on the pocket and attached to the monitor. The shape is fun yet serious, approachable and simple. Neither too playful nor silly, the cylindrical shape is comfortable to hold and visually compelling.

DESIGNERS
Raymond W. Riley,
Ken Weber of Apple Computer, Inc.;
Ken Wood, Mark Edwards of Lunar Design

CLIENT
Apple Computer, Inc.

Sushi Phone

DESIGNER
Robert L. Marvin, Jr. of
Anderson Design Associates

CLIENT
Anderson Design Associates

This cordless telephone handset with a charging base and answering machine uses the metaphor of a shrimp: like phones, shrimp are extremely numerous, much sought after and found on tabletops everywhere. The evocative design gives each function a clearly distinct form. The design's human factors make it easy and comfortable to use.

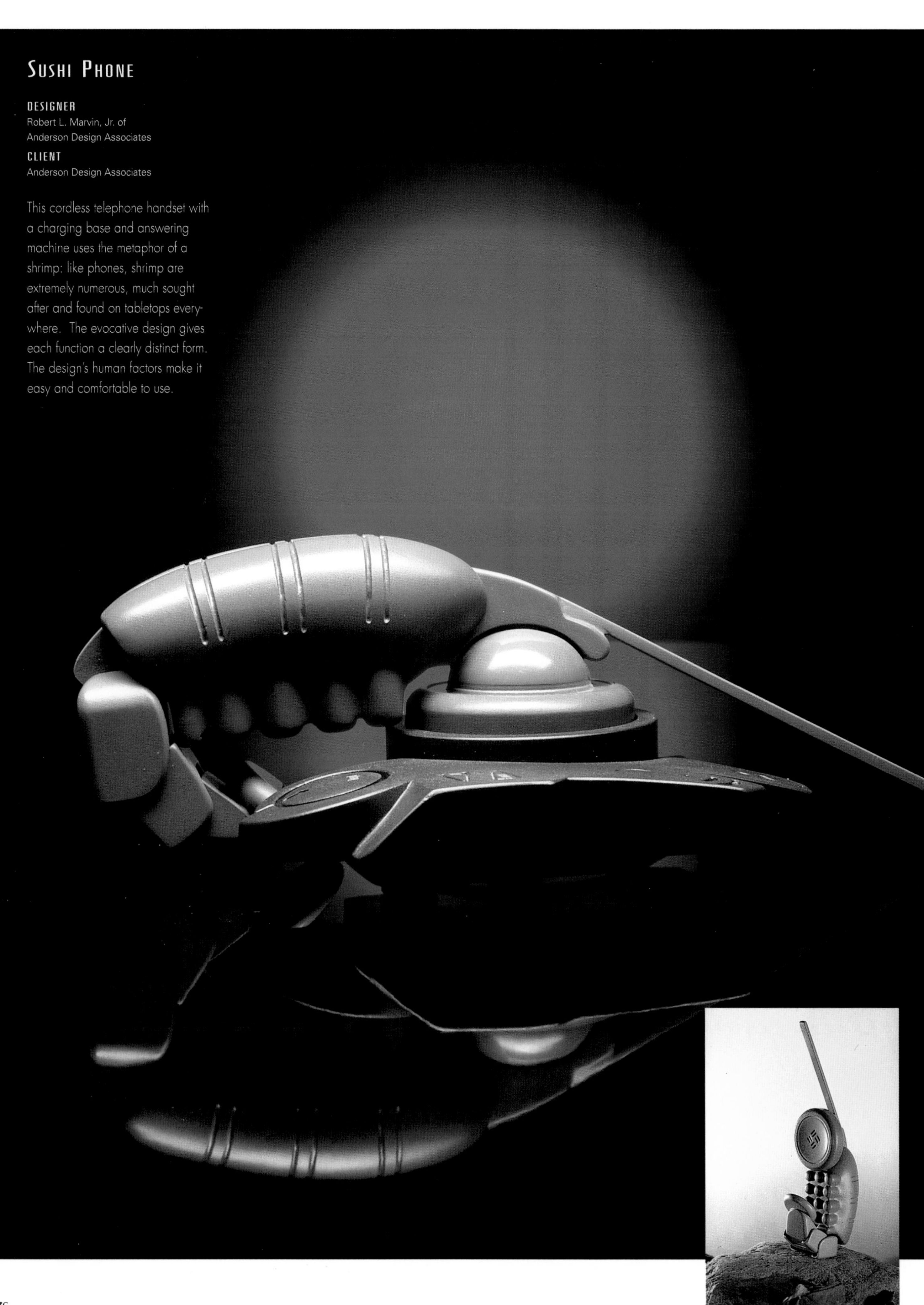

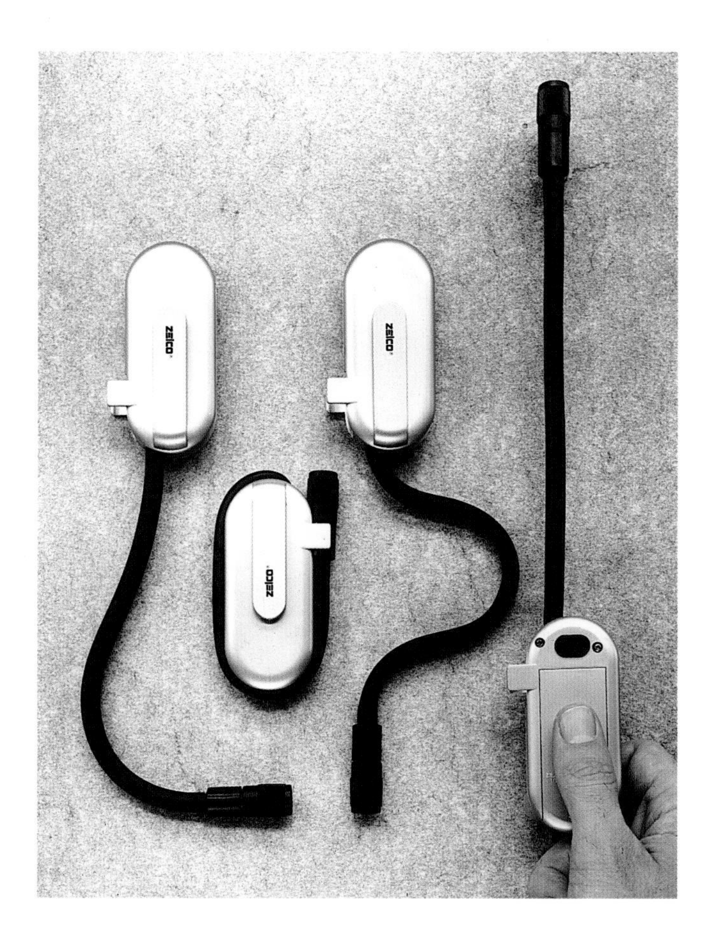

Long Reach Flexible Flashlight

Compact and versatile, this product features a light that can be easily focused from floodlight to spot beam. The unique flexible arm can be bent into any position and stay in place. The arm connects the battery case to the reflector lamp head, which is waterproof. Sales reaction has indicated that more than 300,000 units will have been sold in the first six months of 1991.

DESIGNER
Noel E. Zeller of
Zelco Industries Inc.

CLIENT
Zelco Industries Inc.

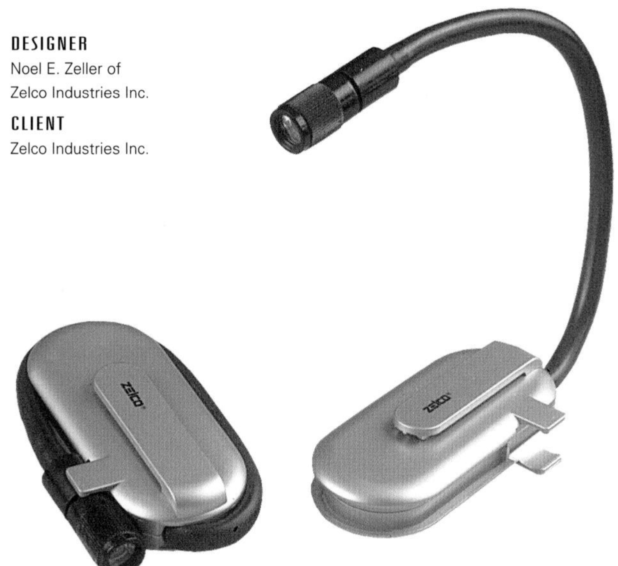

Attiva Seating System

DESIGNER
Jerome Caruso of
Jerome Caruso Design Inc.

CLIENT
Thonet Industries

These chairs provide comfortable, supportive seating for public areas and offices. The one-piece shell responds to the natural flex in the single-piece, wire chair frame, so the frame and seat work together, reacting to even subtle body movements. Because the parts are mass produced with minimal labor time, the Attiva competes with lines produced by larger companies on the basis of both cost and design.

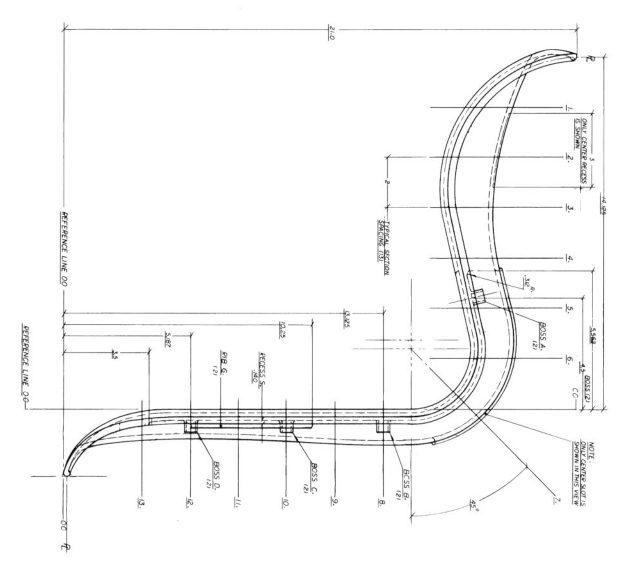

Speedo® SwimMitt™ XT Aquatic Cross Training Gloves

DESIGNERS
Ned Hoffman of
Sports-Mitt International

CLIENT
Speedo

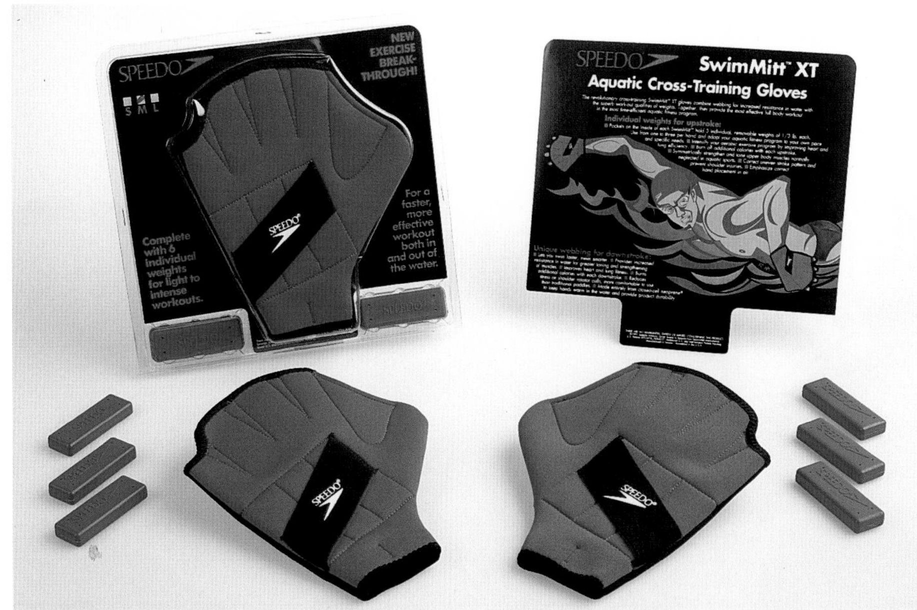

An aquatic cross-training glove, the SwimMitt™ is designed to increase the efficiency of aquatic fitness by intensifying muscular, cardiovascular and pulmonary exercise. Weights are secured in pockets on the back of the hand, and the webbing has an open finger design so users can take their pulse during workouts and retain tactile dexterity. An instant retail success, more than 30,000 pairs were sold in four months.

Sand Soc Footwear

DESIGNERS
Dan Wickemeyer,
David Potter of Slam Design

CLIENT
Reebok USA

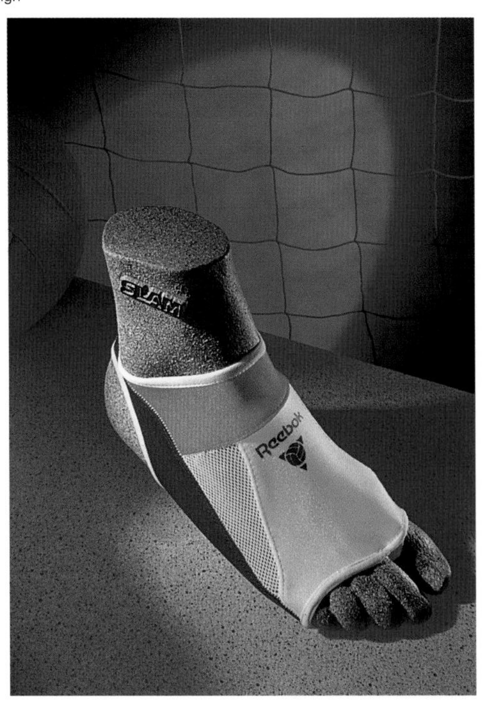

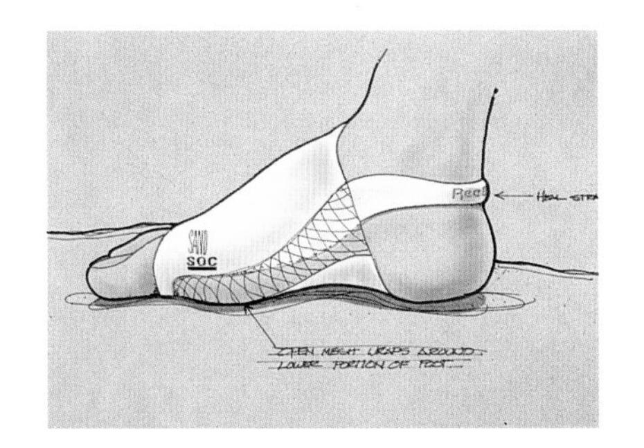

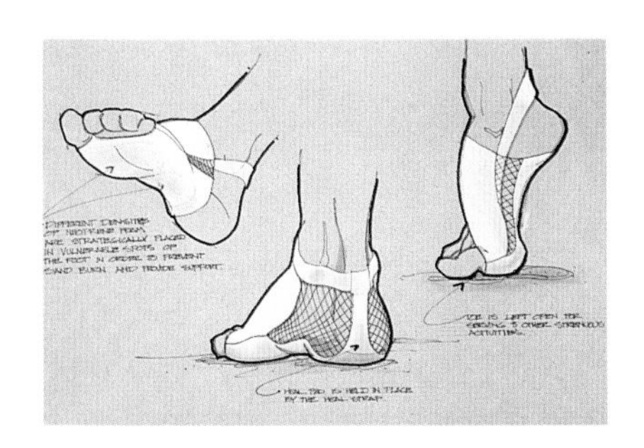

The Sand Soc provides simple, elegant protection for the bare foot while playing volleyball or walking on hot summer sand. With an organic configuration produced in appropriate synthetic fabrics, its design allows the movement of the foot to expel any sand that enters. Human factors were carefully considered in the placement of "cat paw" support pads that are embossed on the sole of the Sand Soc.

Mia Breast Pump

DESIGNERS
Thomas Shoda, Dallas Grove of
Palo Alto Design Group, Inc.

A hand-held, manual breast pump, Mia gives the user control of the suction, pumping cycle and position. The result is a more comfortable and effective process. Unlike existing piston and cylinder-type pumps, which require two hands and high force to extend the piston, the Mia can be used with one hand. The milk feeds directly into standard bottles.

SwingArm Typing Aid

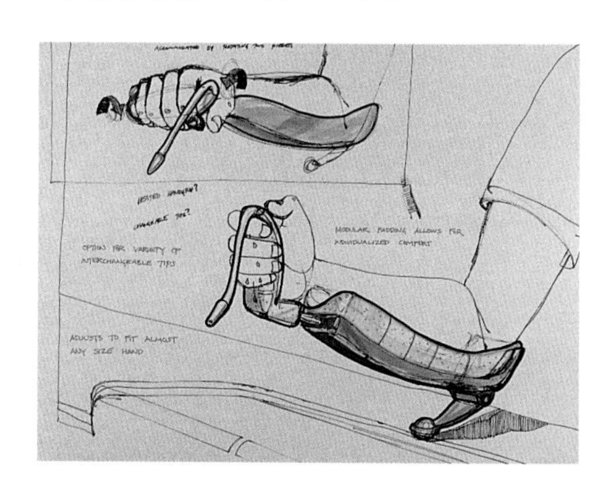

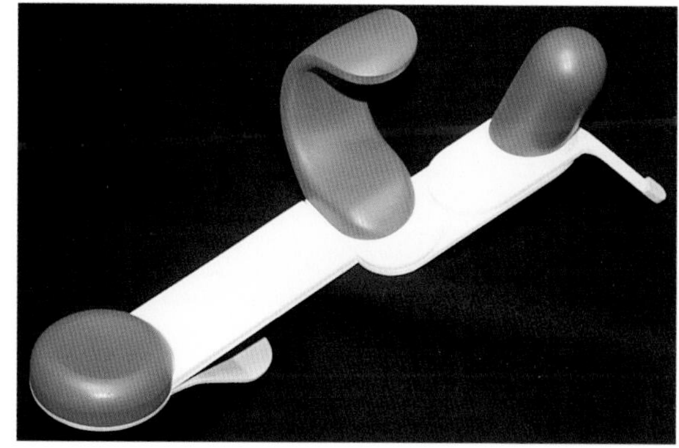

This design allows people who suffer from pain and limited hand movement to use a personal computer. The typing tip pivots right/left and up/down as the elbow rotates. The user's forearm slips under the wrist piece and the hand rests lightly in a variety of positions. A soft foam pad underneath the handpiece allows the user to manipulate a mouse, important with today's graphical computer interfaces.

DESIGNER
Steven M. Slaton

DESIGN SCHOOL
North Carolina State University

SPONSOR
IBM Design Center

Tristander

DESIGNERS
Elsie Bussey, Kris Wohnsen of
Bissell Healthcare Corp.;
Sam Camardello of Tumble Forms Inc.

CLIENT
Bissell Healthcare Corp.

The Tristander provides standing, weight-bearing therapy in three different positions—prone, supine or upright—for children who cannot stand on their own. Inobtrusive and colorfully inviting, its molded foam support pads combine to offer the child comfort and security without being restrictive.

GunSafe

DESIGNERS
Martin Smith, J.B. Kit Dicarlo

Developed to prevent children from accidentally discharging a gun, a full-coverage plastic sheath isolates the trigger. A PVC-coated metal strap fits under the thumb rest area of the hammer, preventing anyone from cocking it. A multiple-tooth detail in the strap adjusts in 1/16-inch increments, allowing the user to fit the GunSafe to their firearm. The design prevents unauthorized opening, but the gun's owner can quickly remove the product in low light and under high stress, unlike a key-based lock.

Woodzig Power Pruner

DESIGNERS
Sohrab Vossoughi, Robert Dillon, David Knaub of ZIBA Design Inc.

CLIENT
Blount Corp.

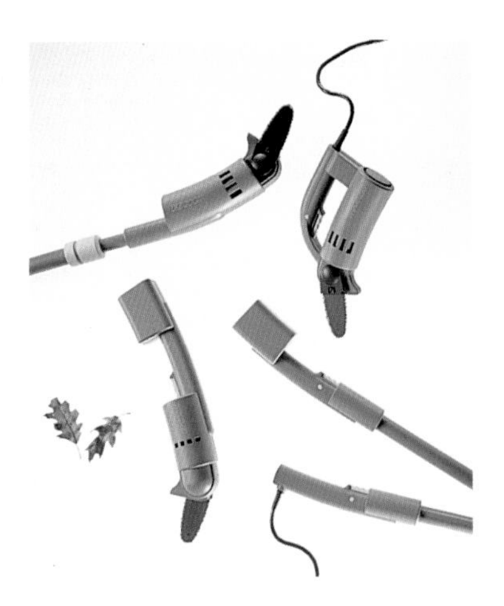

This hand-held pruning tool is the result of exhaustive ergonomic research to optimize the placement and sizing of the grip for comfort across a broad range of users. The grip design also functions as a hand shield. By incorporating fastener details in the plastic housing, the design reduces part count and assembly time. Moreover, roughly 80 percent of the parts are leveraged across all four versions of the pruners, producing significant savings.

Encore

DESIGNERS
Sohrab Vossoughi,
Tom Froning of ZIBA Design Inc.
CLIENT
Cadet Manufacturing Co.

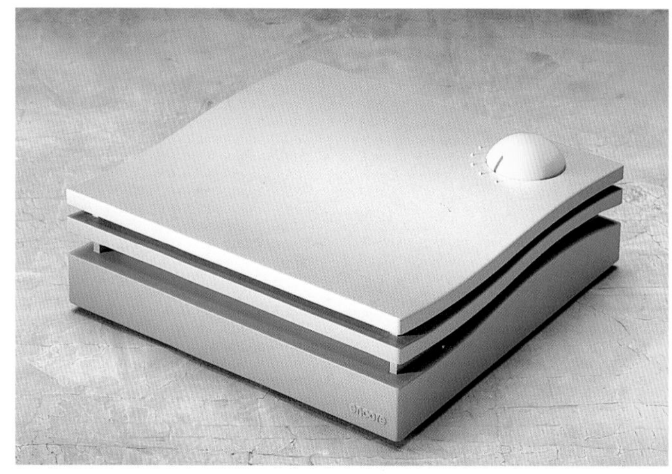

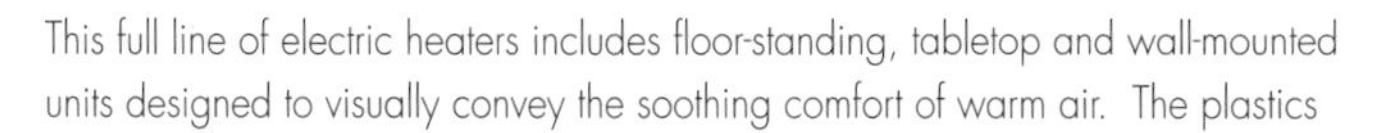
This full line of electric heaters includes floor-standing, tabletop and wall-mounted units designed to visually convey the soothing comfort of warm air. The plastics are identified for easy recycling and, with only one screw in each heater, the design facilitates disassembly.

One-Piece Fishing Pliers

DESIGNERS
Steve Visser,
Miro Tasic,
Ashok Midha

This concept for a convenient fish hook remover is based on a compliant mechanism—one part flexes while the rest remains rigid. The single-piece design makes the pliers easy to manufacture and recycle. Injection molded of a durable structural plastic—Delrin—the device will not corrode, and will float if accidentally dropped overboard.

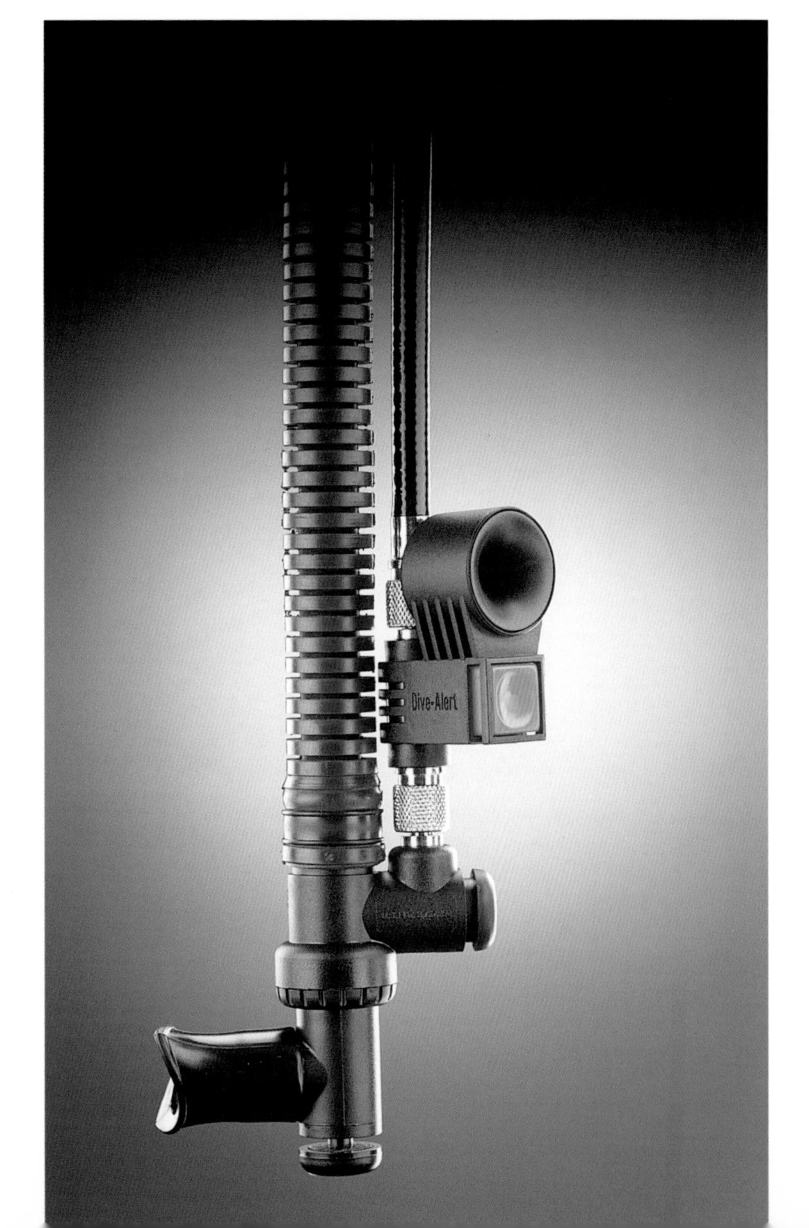

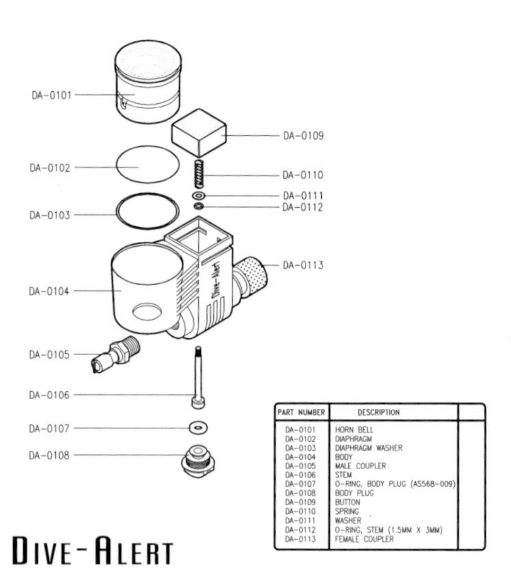

PART NUMBER	DESCRIPTION	
DA-0101	HORN BELL	
DA-0102	DIAPHRAGM	
DA-0103	DIAPHRAGM WASHER	
DA-0104	BODY	
DA-0105	MALE COUPLER	
DA-0106	STEM	
DA-0107	O-RING, BODY PLUG (AS568-009)	
DA-0108	BODY PLUG	
DA-0109	BUTTON	
DA-0110	SPRING	
DA-0111	WASHER	
DA-0112	O-RING, STEM (1.5MM X 3MM)	
DA-0113	FEMALE COUPLER	

Dive-Alert

DESIGNERS
Dave Hancock, Barry Kornett of Ideations, design inc.

CLIENT
Ideations, design inc.

Dive-Alert is a signaling device for scuba divers in trouble or lost at sea. At the mere touch of a button, Dive-Alert generates a blast of sound that can be heard up to a mile away. The design taps into the energy source that every diver already carries—the air tank—and integrates with the diver's existing equipment without modification. It is currently sold in more than 500 stores in the US and in 30 countries.

Cachet Articulated Stamp

DESIGNERS
David Mehaffey, James Machen,
Ed Boyd, Eddie Machen,
Edgar Montague of Machen Montague

CLIENT
US Stamp

This design eliminates the need for a separate dust cover and ink pad sold with most ink stamps. The angled bull nose grip directs the label toward the user. The vertically stored ink platen reduces its footprint and the clear lens over the identification label provides protection and cleanliness.

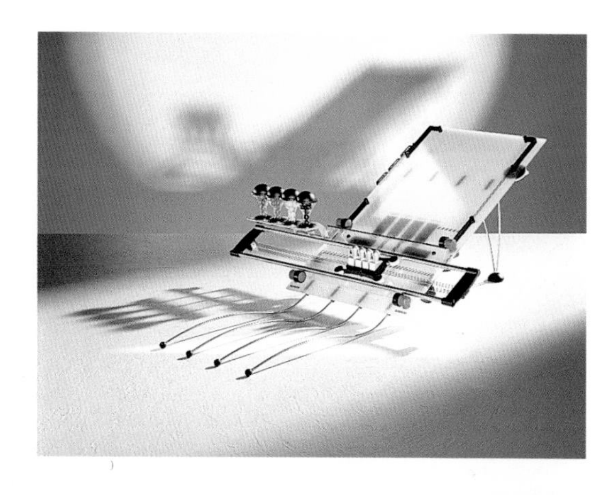

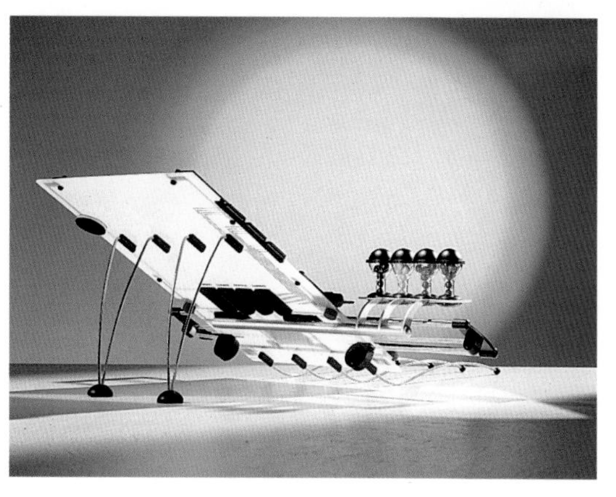

Ithaca Solid-Ink Color Printer

DESIGNER
Donald Carr

DESIGN SCHOOL
Cranbrook Academy of Art

SPONSOR
NCR Corp.

Solid-ink color printers are in essence the fusion of fluids and electronic input. Where the two merge, an image is created. Ink, embedded in a wax carrier, is heated within the print-head and jetted onto the paper. There is an inherent beauty in being able to view this phenomenon. This design externalizes the primary components of the printer, with informational cues regarding its status prominently displayed.

Finishing Planer

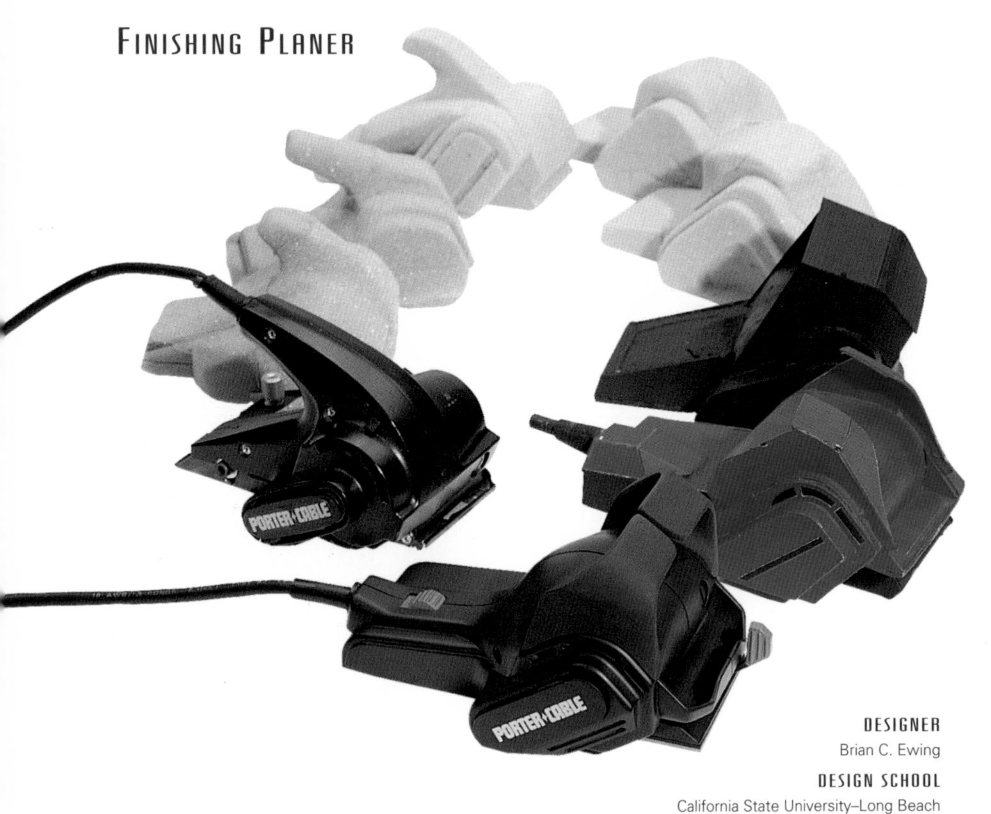

This product analysis and redesign includes a complete cost breakdown and results in a much safer design. Noting that planers are often held incorrectly, causing injury, the designer produced an ambidextrous and more comfortable hand grip. A feather blade height adjustment enables the user to vary the blade height while the product is in use simply by sliding the front red lever from left to right or vice versa.

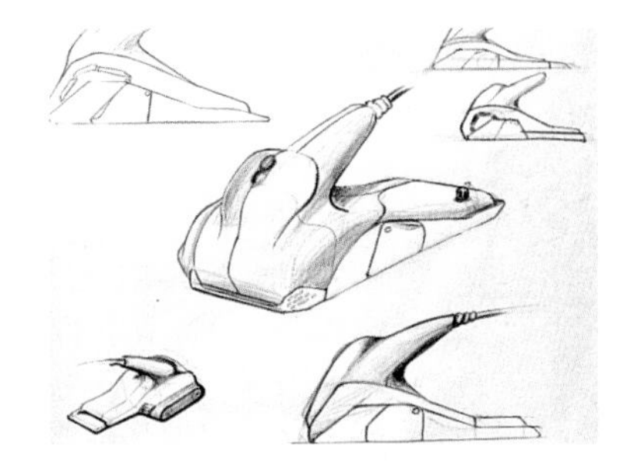

DESIGNER
Brian C. Ewing

DESIGN SCHOOL
California State University–Long Beach

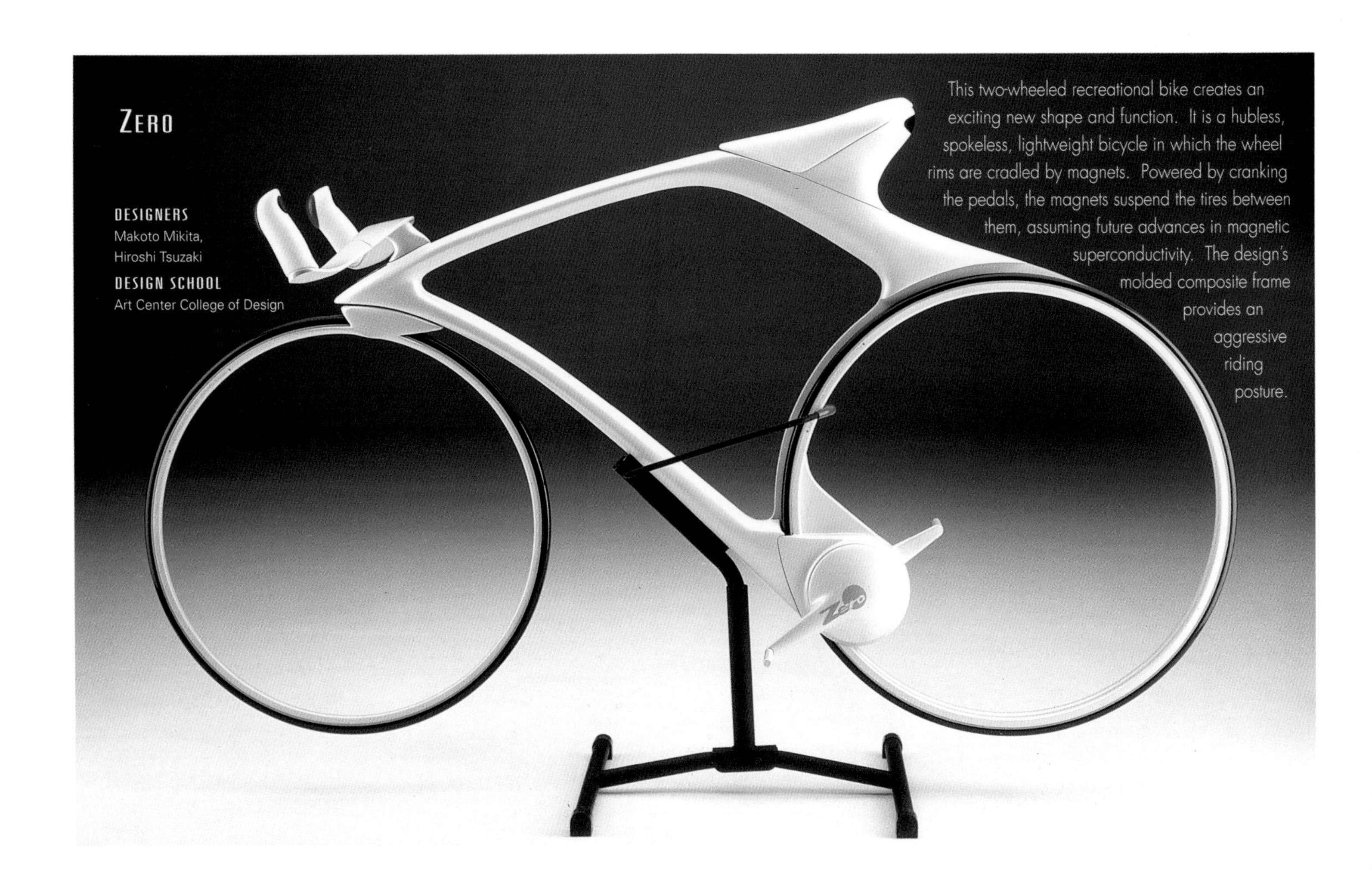

Zero

DESIGNERS
Makoto Mikita,
Hiroshi Tsuzaki

DESIGN SCHOOL
Art Center College of Design

This two-wheeled recreational bike creates an exciting new shape and function. It is a hubless, spokeless, lightweight bicycle in which the wheel rims are cradled by magnets. Powered by cranking the pedals, the magnets suspend the tires between them, assuming future advances in magnetic superconductivity. The design's molded composite frame provides an aggressive riding posture.

Crime Shield Window Barriers

DESIGNERS
Robert J. Cohn, Albert Kolvites of Product Solutions Inc.; Ed Kelleher of Exeter Architectural Products

CLIENT
Exeter Architectural Products

Crime Shield protects against entry through a window with a tough, perforated steel membrane. The framing system affords continuous protection around the full perimeter of the window without the exposed fastening points of competitive devices. The single-point, quick-release latch and swing-open barrier allows fast escape in an emergency, but tamper seals effectively discourage unauthorized opening otherwise. Sales have surpassed all original expectations and market penetration has been quick and deep.

Collection Fleury Stacking Chair

DESIGNERS
Thomas Gerlach, Gabor Lengyel of frogdesign inc.

CLIENT
Jardin

This garden stacking chair, made of recyclable polypropylene plastic, achieves a recognizable image in a market of low-priced, look-alike competitors. The shape has the quality of good indoor furniture with office chair comfort.

Io Positional Track Lighting

DESIGNERS
Tim C. Repp, Richard Barbera

This design addresses the user interface problems of track lighting, which needs adjustment to suit varying displays in stores and museums. A simple wand that clips onto the light eliminates the need to climb an 8- to 12-foot ladder in order to aim the light source.

Rage Jet Boat

DESIGNERS
Peter J. Van Lancker, Scott Wood, Rick Strand of Boston Whaler, Inc.

CLIENT
Boston Whaler, Inc.

The design uses an existing jet ski drive train. With no propeller, the bottom of the boat is completely flush. This boat is unsinkable and nearly impossible to flip: at a sudden hard overturn even at full speed (33 mph), the boat stays level. Moreover, the boat's clean passage minimally disturbs the environment.

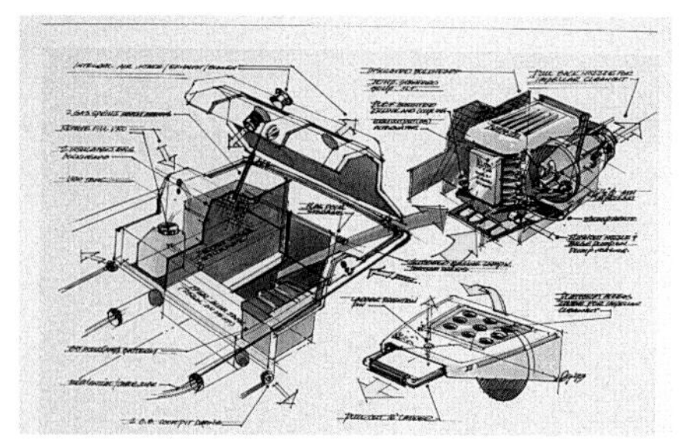

BodyBoat

DESIGNER
Donna Cohn

DESIGN SCHOOL
Rhode Island School of Design

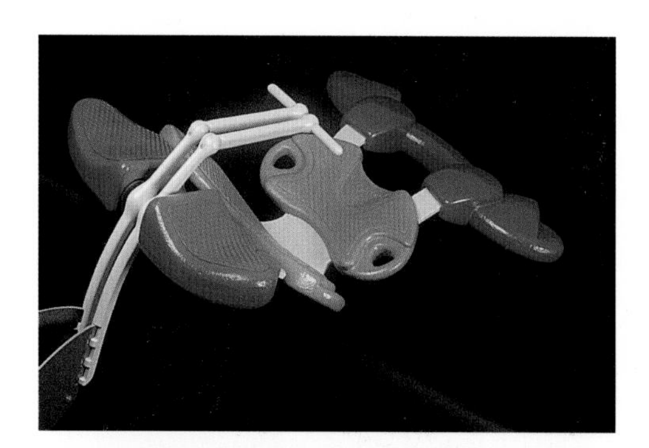

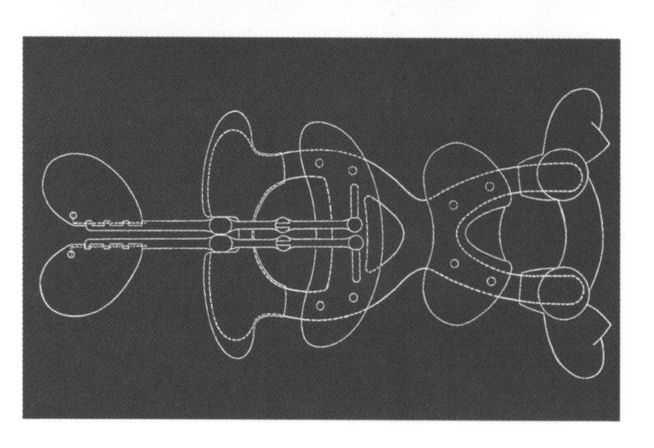

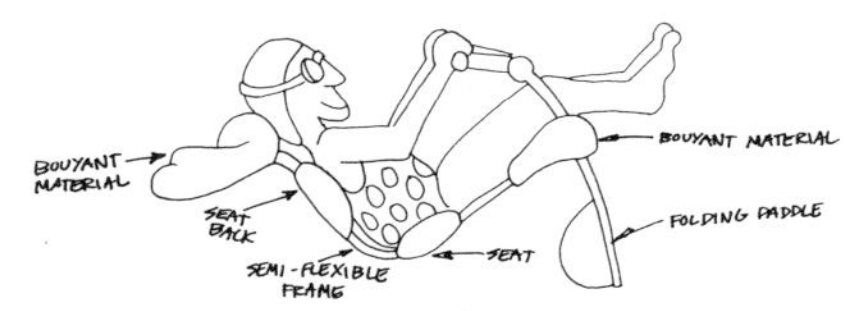

BodyBoat is a floatation and propulsion device for physically disabled children who have some use of their arms. They move the boat through the water by pumping a lever. The BodyBoat keeps the user's head and upper torso above water, while supporting his or her body below the water's surface in a semi-reclined position. The form is playful, bright and inviting to any child.

Casting Reel

DESIGNERS
Sohrab Vossoughi,
Mark Stella, Ken Dieringer and
Kuni Masuda of ZIBA Design;
Vic Culler of Fenwick

CLIENT
Fenwick

This lightweight reel for the spin fisher demonstrates outstanding hand-held ergonomics and visual styling. An existing OEM reel mechanism was positioned off-center to provide a more comfortable and better controlled cast. The reel competes in the high-end market against reels costing twice as much.

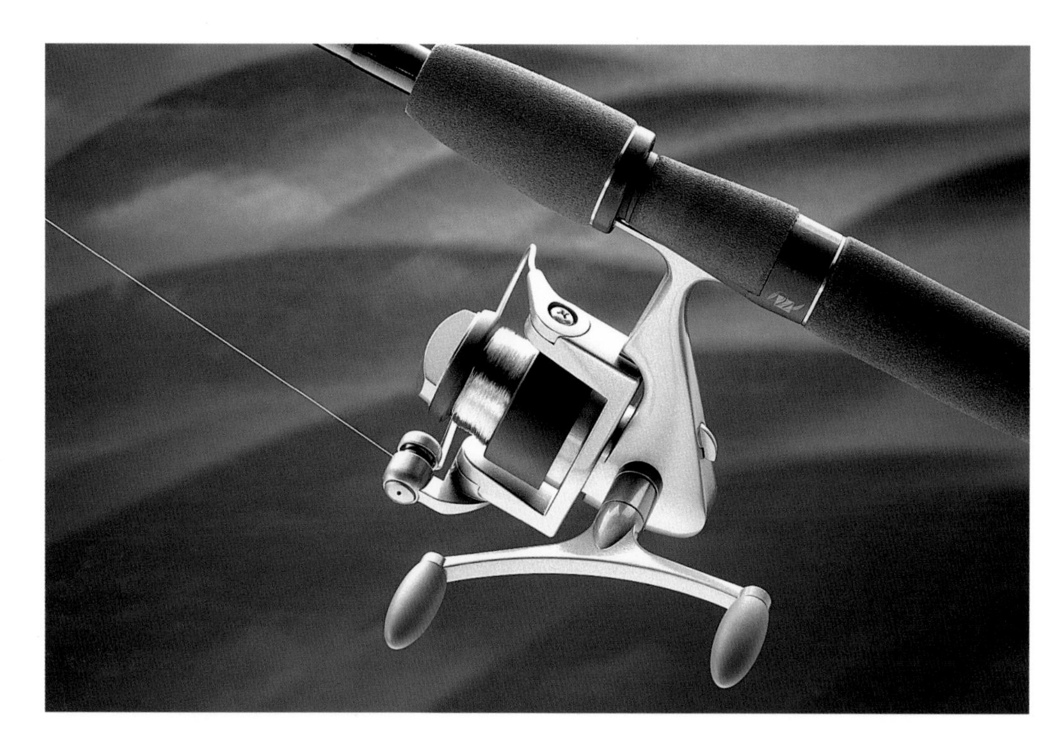

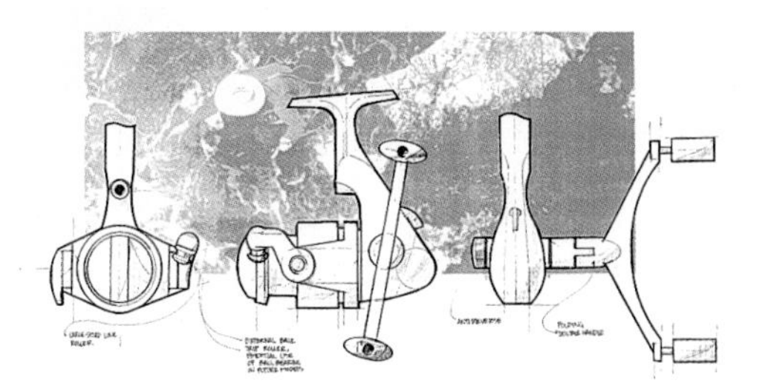

Leaf Eater

DESIGNERS
Edwin Beck of Fitch, Inc.;
Chris Cicenas of Polymer Solutions, Inc.;
Sal DeYoreo of Flowtron Outdoor Products

CLIENT
Flowtron

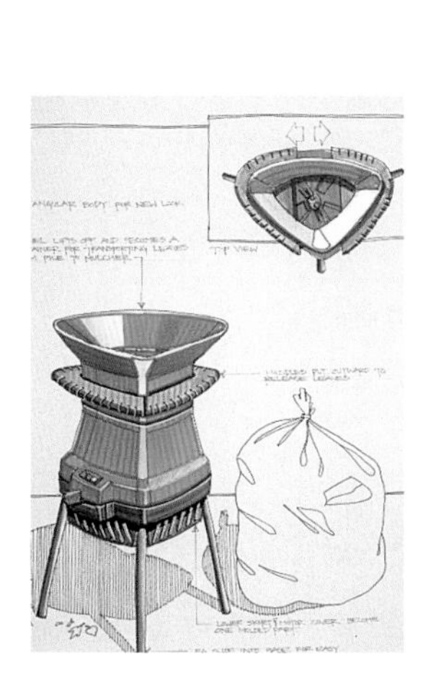

This electric lawn debris and leaf munching device increases the amount of leaves that can be put in one bag by 800 percent. A redesign, the new Leaf Eater is perceived as more performance-oriented and durable, featuring a rigid funnel, a loop leg in place of the wobbly tripod legs, and a fortified safety guard. The clam shell and snap-fit design reduced tooling costs and assembly time.

Metroblade

An in-line rollerskate, Metroblade includes a detachable shoe, so that skaters who use the Metroblade as transportation only have to remove the skate upon arrival rather than carrying shoes to change into. Simple to use, the design is lightweight, supportive and ventilated, with a padded elastic construction around the ankle to prevent pinching.

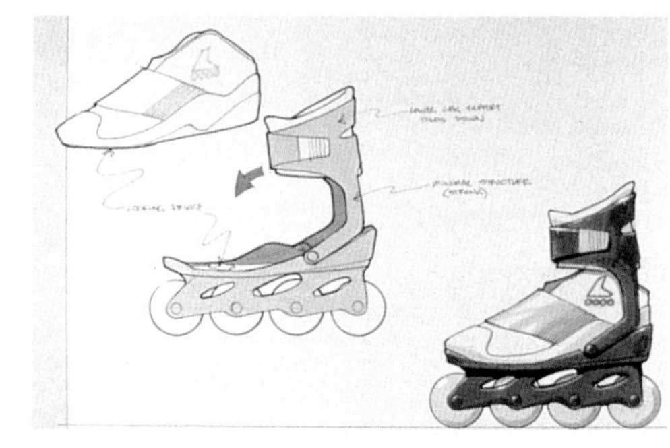

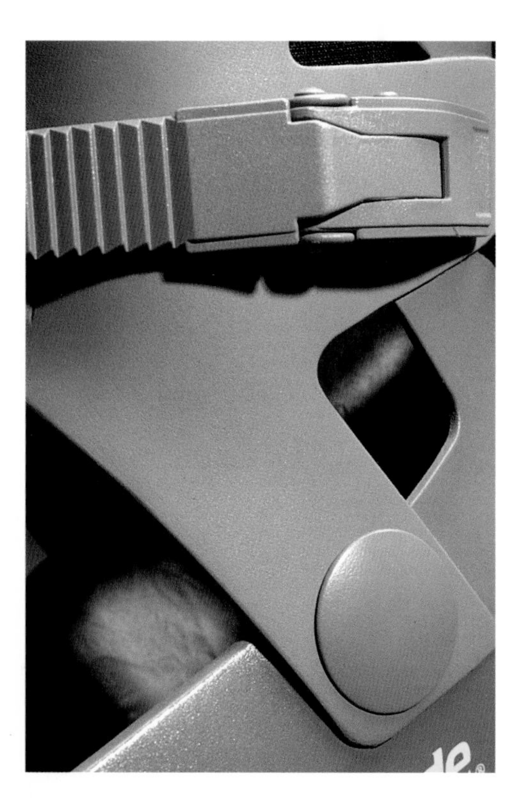

DESIGNERS

Michel Arney, Carl Madore, Andrew Jones, Harvey Koselka of Design Continuum Inc.; Jack Curley of Jen Jen International

CLIENT

Rollerblade Inc.

Römer "frogdesign one"

This full-mask motorcycle helmet is put on by sliding the chin section forward. The back integrates a venting system that creates a vacuum-effect because of its form. Safer and more comfortable than other helmets, this innovative product has caused sales to triple.

DESIGNER
the frogdesign team

CLIENT
Römer GmbH

Orion AW Photography Backpack

DESIGNER
Doug Murdoch of
Resource International

CLIENT
Lowe Pro International

Used by photographers who work outdoors, this lightweight, waterproof design consists of a day pack and a fanny pack. The fanny pack, with its padded waistband, gives easy access to the camera equipment through a novel lid that opens away from the user. Contoured to fit the user's back and to allow free movement of the arms, the day pack is used primarily to hold water, food and clothing.

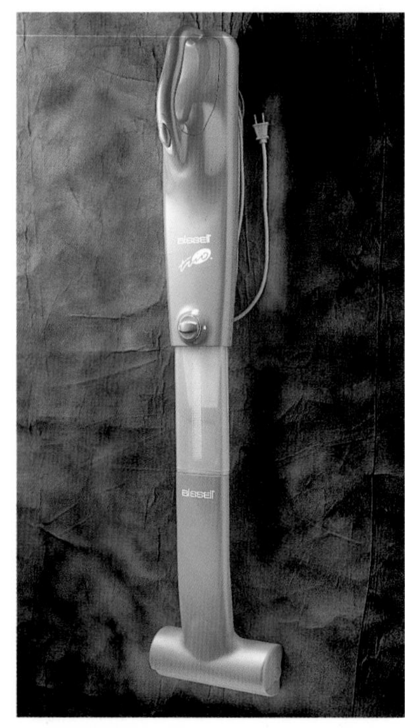

Trio Vacuum

DESIGNERS
Bissell Inc. Industrial Design; INNO

CLIENT
Bissell Homecare Div.

The Trio easily converts between a hand-held vacuum and powered "stick" broom. The multiple position handle and lightweight design facilitates use by a large range of users for many tasks. Tinted semitranslucent polycarbonate used for the dirt receptacle allows the user to "see" when the unit needs to be emptied.

T-245 Metropolitan Toaster

DESIGNERS
Gary Van Deursen, Don McCloskey,
Stuart Naft of Black & Decker;
Tucker Viemeister of Smart Design Inc.

CLIENT
Black & Decker
US Household Products Group

This toaster looks back to the softer forms from the American kitchens of the past. The design reduces the number of external components by 60 percent, providing cost and ecological benefits. Polypropylene replaces the typical polycarbonate material used for the housing to provide a lower cost to the consumer and higher profit to the manufacturer. The exterior stays cool during toasting and is easier to clean than the standard toaster.

Convertible Pen-Notebook Computer

DESIGNERS
Mark Biasotti of
IDEO Product Development;
Jack Daley, Jeff Hawkins of
Grid Systems Corp.

CLIENT
Grid Systems Corp.

This convertible portable notebook computer provides users with a pen-based interface that seamlessly converts to a keyboard-based system with a flip of the screen. The integration of both inputting methods eliminates the need for customers to have to choose between the types of portable computers available, since different tasks require different manners of working.

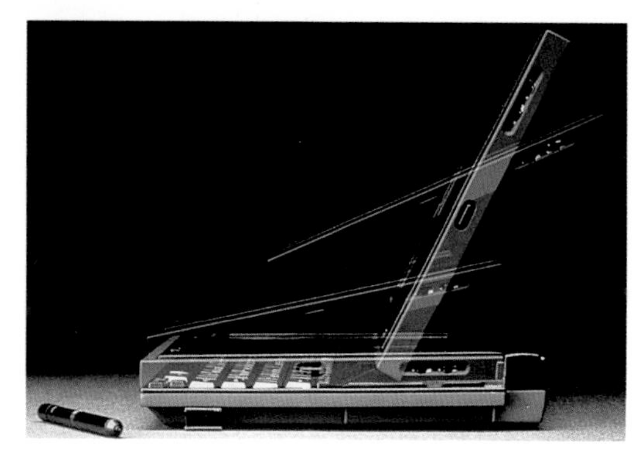

Spinal Cord Injury Patient Prone Cart

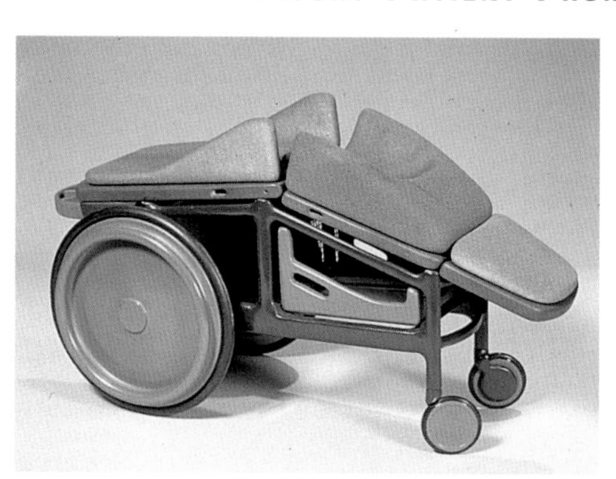

DESIGNERS
Todd Hoehn, Robert Meurer

FACULTY ADVISORS
Pascal Malassigné, Brooks Stevens

DESIGN SCHOOL
Milwaukee Institute of Art & Design

SPONSOR
The CJ Zablocki VA Medical Center

Designed for use by spinal cord injury patients, this cart offers increased mobility. Current designs do not offer victims the chance to interact with their peers and often isolate them from society. This design offers adjustable height through hydraulic lifts and articulation of the front and rear panels. The design includes trays for writing and eating, and baskets for storage.

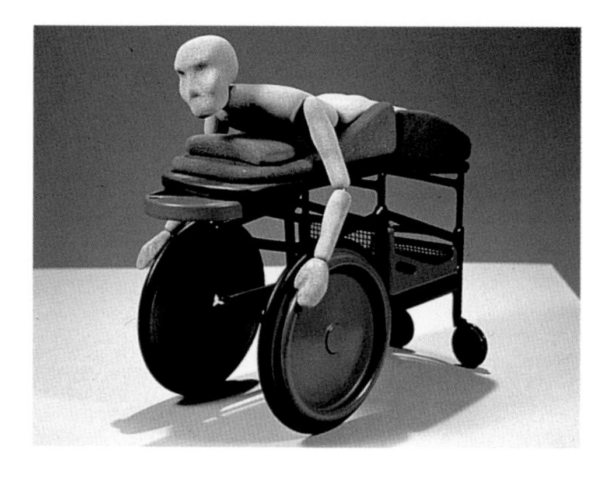

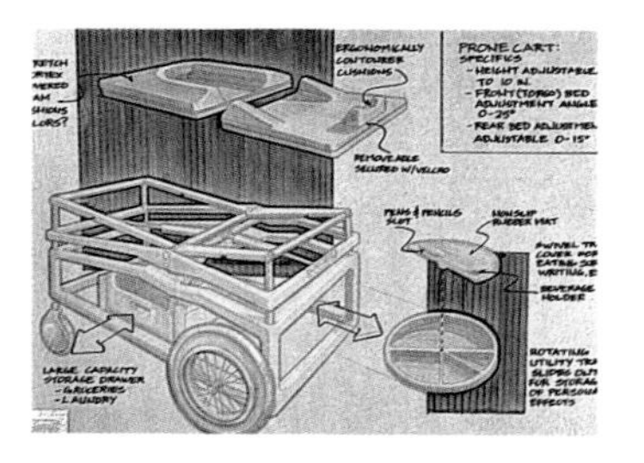

Portable Presentation System

DESIGNER
Derek Gratz

DESIGN SCHOOL
Art Center College of Design

SPONSOR
Steelcase

This system provides the traveling professional a high quality image. The collapsible table and tamboured large screen are peripheral units for use when a desk is needed or a larger audience demands a bigger viewing area. All presentation information is visually formatted on the laptop briefcase unit.

Orbit Lawnmower

DESIGNER
Daniel R. Vehse

FACULTY ADVISOR
Raymond Smith

DESIGN SCHOOL
Art Center College of Design

SPONSOR
Honda Power Equipment

The Orbit concept features a mulching system, centrally located control panel, unique center wheel, gimbaled blade, plastic housing, an adjustable height handle, and battery-powered operation.

Quest

DESIGNERS
Tom Semple, Bruce Campbell,
Diane Taraskavage, Gerald P. Hirshberg of
Nissan Design International

CLIENT
Nissan

The Quest, a front-wheel-drive minivan, was designed as a "roomy, flexible, comfortable interior volume on wheels," according to Nissan. A unique sliding/folding seat system facilitates easy conversion from passenger to cargo van. The van meets the 1993 safety requirements set for cars which are more stringent than those governing vans.

Probe GT

DESIGNERS
Toshi Saito, Mimi Vandermolen,
Mark Kelly, Richard Chung,
Syd Chiang of Ford Motor Co.
CLIENT
Ford Motor Co.

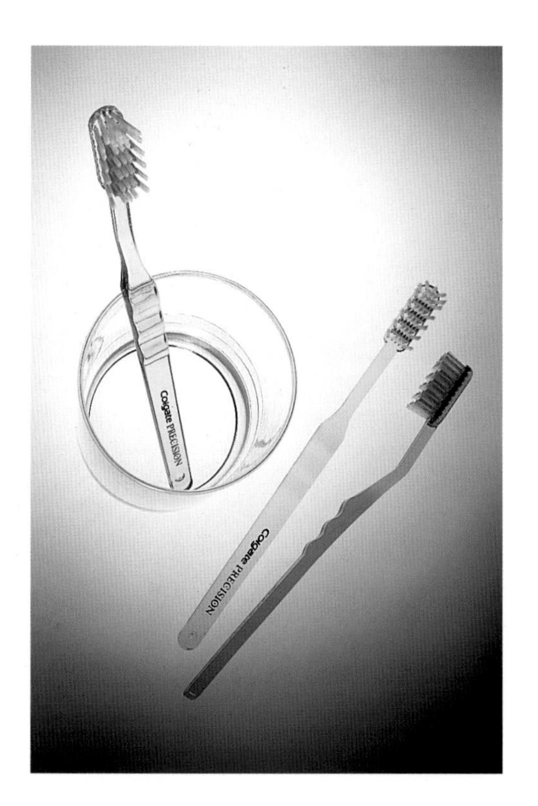

Precision Toothbrush

DESIGNERS
Bert Heinzelman, Don Lamond,
Laura Edelman, Douglas Spranger of
Human Factors Industrial Design, Inc.;
Jay Crawford of Colgate
CLIENT
Colgate-Palmolive

Yontech 105

DESIGNERS
Alfonso Albaisa, Gerald Hirshberg, Allan Flowers, Bruce Campbell of Nissan Design International, Inc.

CLIENT
Yonca Teknik Yatirim AS

Jeep

DESIGNER
Michael Kent

DESIGN SCHOOL
Art Center College of Design

SPONSOR
Chrysler Corp.

DAL Fixtures

DESIGNER
the frogdesign team

CLIENT
DAL

Cruiser/Commuter

DESIGNER
Al Arrosagaray

FACULTY ADVISORS
Imre Molnar, Harry Bradley, Tom Campbell

DESIGN SCHOOL
Art Center College of Design

Trackman II

DESIGNERS
Tino Melzer, Dan Harden of frogdesign inc.

CLIENT
Logitech, Inc.

Orbital Finishing Sander

DESIGNER
Greg Stuhl

FACULTY ADVISOR
Michael Kammermeyer

DESIGN SCHOOL
California State University–Long Beach

PalmPAD

DESIGNERS
Nelson Au of IDEO Product Development;
Dennis Silva, Kate Purmal of GRiD Systems Corp.

CLIENT
GRiD Systems Corp.

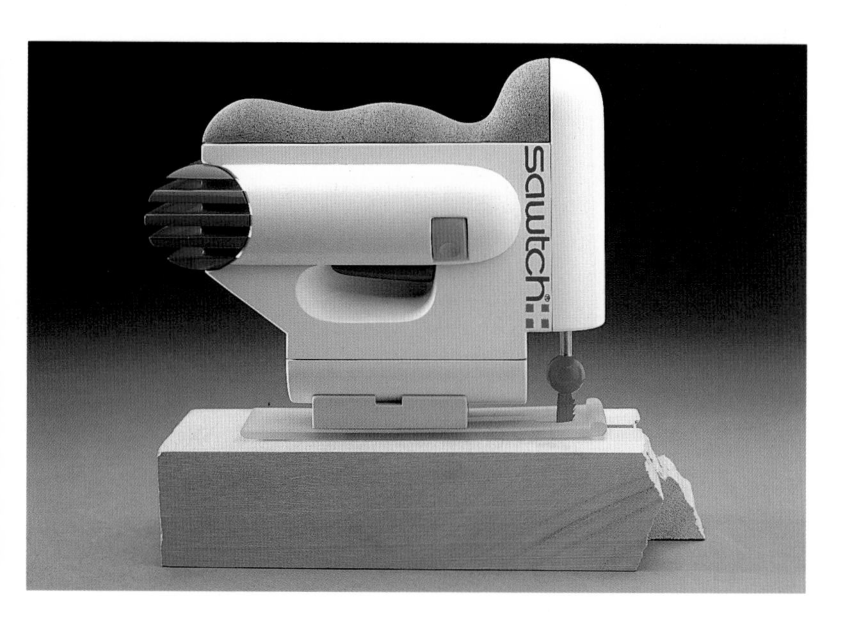

Sawtch Reciprocating Saw

DESIGNER
Holger Schubert

DESIGN SCHOOL
Art Center College of Design

StyleWriter Inkjet Printer

DESIGNERS
Mark Pruitt of Apple Computer, Inc.;
Doug Patton, Rick Jung, Dennis Grudt of
Patton Design Enterprises

CLIENT
Apple Computer, Inc.

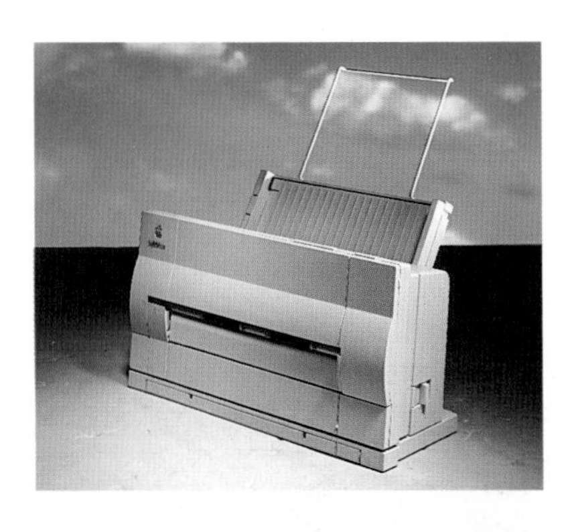

YST-SD90 Audio Speaker

DESIGNER
Hiroaki Kozu of Kozu Design

CLIENT
Yamaha Corp.

Trigem 386SX Laptop Computer

DESIGNERS
Bill Evans, Max Yoshimoto,
Robert Brunner, Gil Wong, Braxton Lathrop,
Keith Willows, Gerard Furbershaw,
Marieke Van Wijnen of Lunar Design;
Sung Kim, Gregor Berkowitz of
Function Engineering;
Tae Lee of Trigem Corp.

CLIENT
Trigem Corp.

Quantum

DESIGNERS
William King of Samsonite Corp.;
Dana Franklin of DANACO

CLIENT
Samsonite Corp.

LaserWriter Pro 600 & Pro 630

DESIGNERS
Ken Wood, Jeff Smith of Lunar Design;
Robert Brunner, Jim Stewart,
Grant Ross, Jr. of Apple Computer, Inc.

CLIENT
Apple Computer, Inc.

Portable Computer Product Family

DESIGNERS
Mark S. Kimbrough,
Richard Haner of Design Edge

CLIENT
Dell Computer

Helper

DESIGNERS
Robert L. Marvin, Jr., David W. Kaiser of Anderson Design Associates

CLIENT
Anderson Design Associates

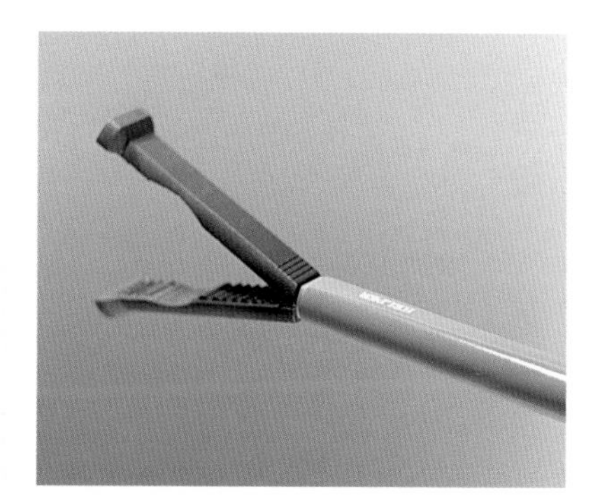

Garment Bag Plus

DESIGNER
James Fournier of JLF Designs Inc.

CLIENT
Bennington Leather

Corded Mouseman

DESIGNERS
Hartmut Esslinger, Dan Harden of frogdesign inc.

CLIENT
Logitech Inc.

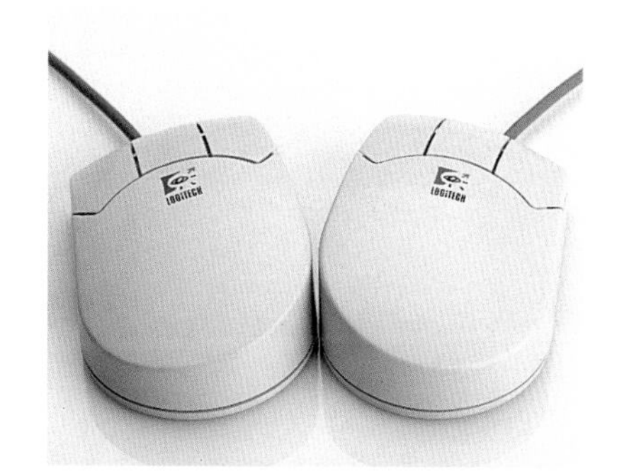

St. Tropez

DESIGNERS
Colin Burns, Bob Yuan of ID TWO; Douglas Warner of Levolor Corp.

CLIENT
Levolor Corp.

Gig Saw

DESIGNER
Frank Sterpka

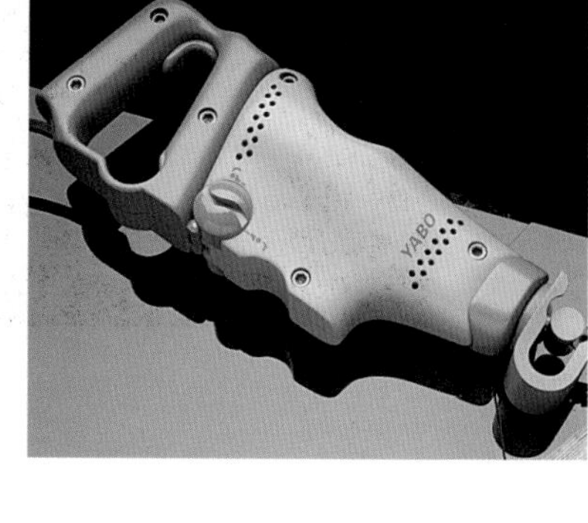

Tucker Recycling System

DESIGNERS
David Evans, Frank DiSesa of Tucker Housewares

CLIENT
Tucker Housewares

Handi-Scratch Wire Brush

DESIGNERS
Sean Simmons, Monte Levin of
Sonneman Design Group Inc.

CLIENT
Empire Brushes Inc.

BallPoint Mouse

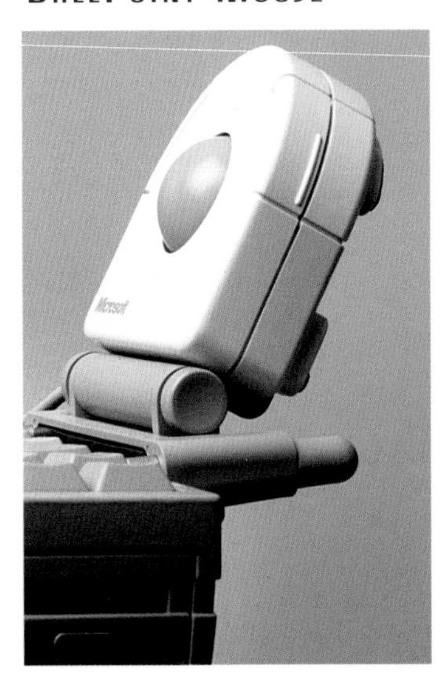

DESIGNERS
Stuart Ashmun of Microsoft Corp.;
Mike Nelson, Mike Paull of
Stratos Product Development;
Paul Bradley of Matrix Product Design

CLIENT
Microsoft Corp.

Tecno Qualis

DESIGNER
Emilio Ambasz of
Emilio Ambasz Design Group

CLIENT
Tecno SpA

HEARSAY

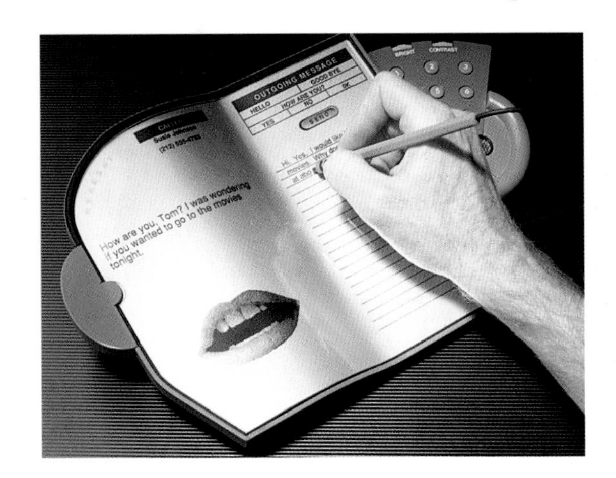

DESIGNER
Barry Sween

FACULTY ADVISOR
Michael Kammermeyer

DESIGN SCHOOL
California State University–Long Beach

Individual Microwave Drip Coffeemaker Concept

DESIGNER
Gary E. Van Deursen of Black & Decker Inc.
CLIENT
Black & Decker Inc.

F20706FT 20-inch Television

DESIGNERS
Thomas E. Renk, Robert Ramspacher of Thomson Consumer Electronics
CLIENT
Thomson Consumer Electronics

Typist Scanner

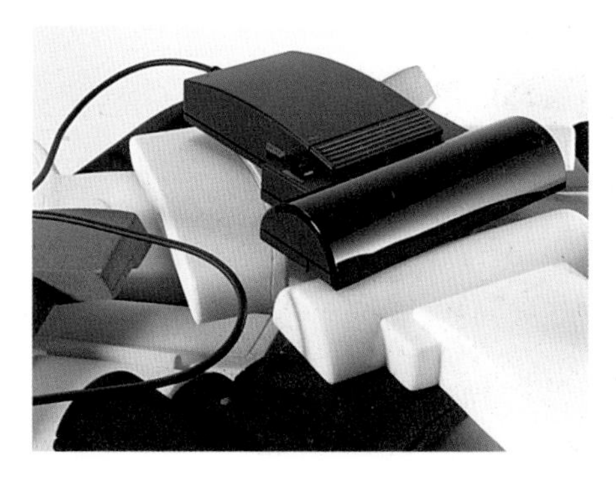

DESIGNERS
Paul Bradley, Tino Melzer of Matrix Product Design Inc.
CLIENT
Caere Corp.

PS/1 Computer

DESIGNERS
Steven Silverstein of IBM Corp.; Richard Sapper
CLIENT
IBM Corp.

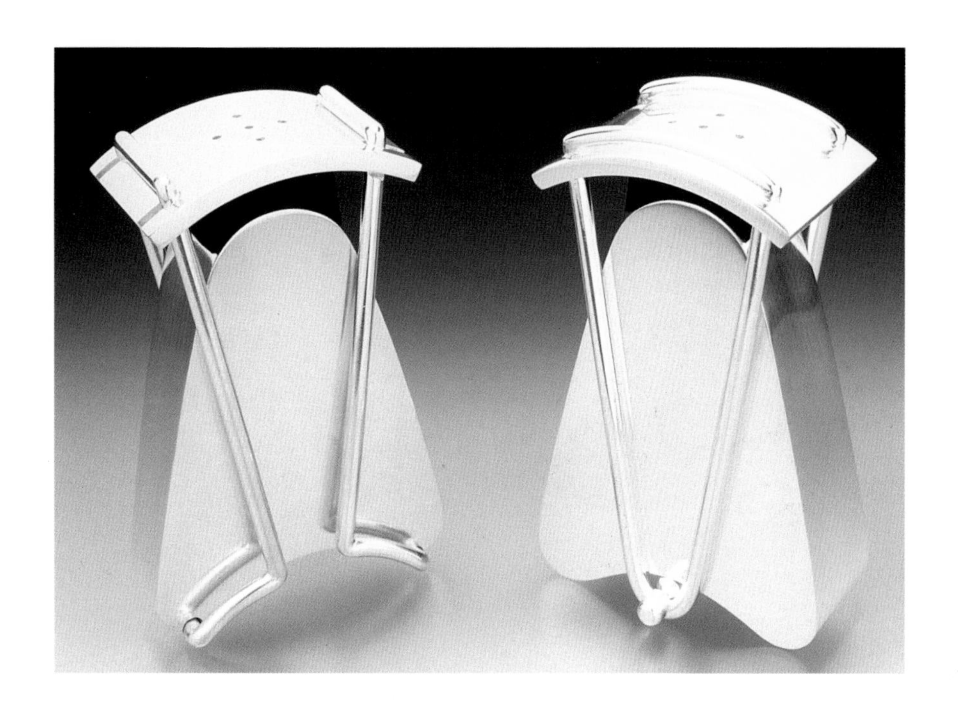

Salt and Pepper Shakers

DESIGNER
Todd Zeilinger

DESIGN SCHOOL
Center for Creative Studies

Electric Moto Cross (EMX)

DESIGNER
William Badsey of Badsey Design of California

CLIENT
Badsey Design of California

Toothbrushes

DESIGNERS
Emilio Ambasz,
David Robinson, Barry Scott of
Emilio Ambasz Design Group

CLIENT
Sunstar, Inc.

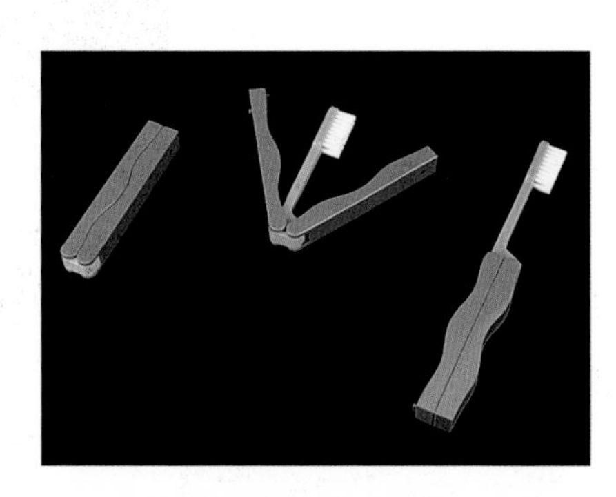

Terrace

DESIGNERS
Anton Dresden, Buzz Siler of
Siler/Siler Ventures

CLIENT
Siler/Siler Ventures

ROLLiter

DESIGNERS
Scott Voorhees of
Inventure Development Corp.;
Tim Parsey

CLIENT
Inventure Development Corp.

Mirage Chaise Longue

DESIGNER
Frederic Doughty of Brown Jordan

CLIENT
Brown Jordan

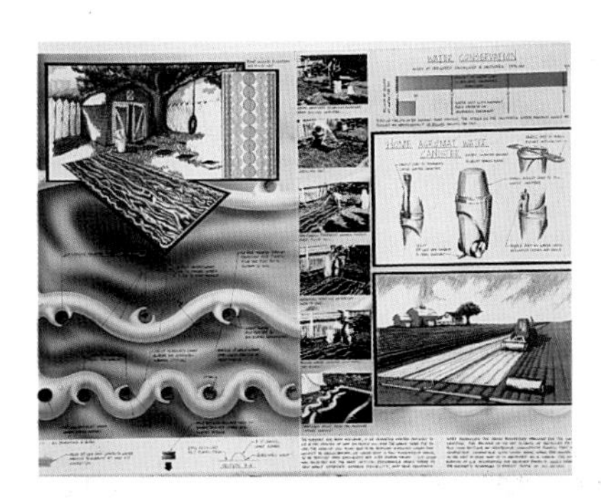

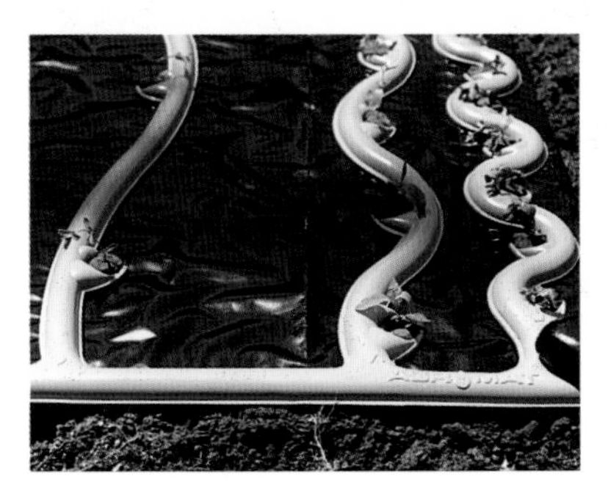

Agromat

DESIGNERS
Paul Best, Dave Butzko, Greg Coyne

FACULTY ADVISORS
Sharyn Thompson, Colin Healy

DESIGN SCHOOL
University of Bridgeport

SPONSOR
General Electric Plastics

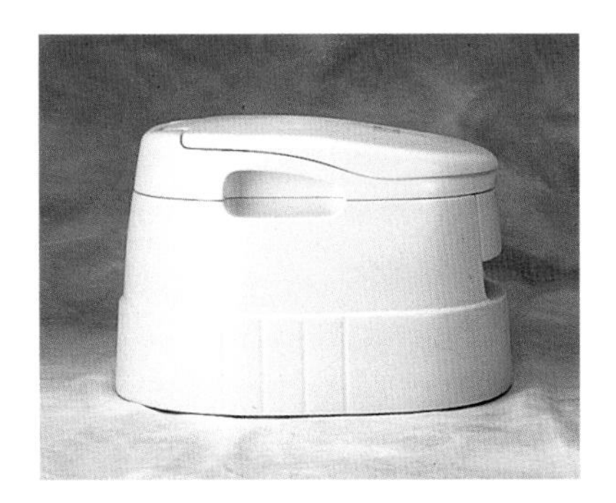

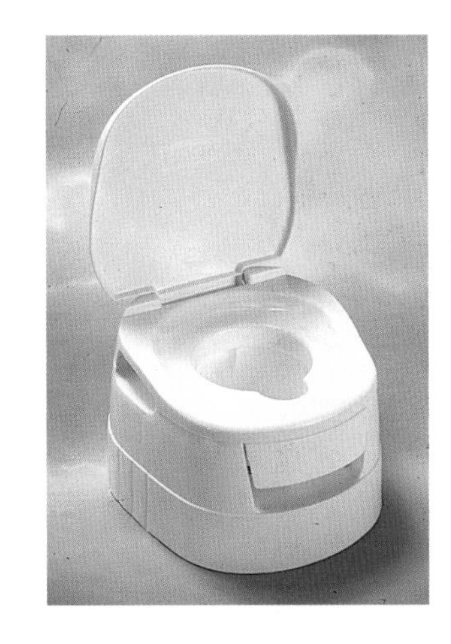

Potty Seat

DESIGNER
Richard M. O'Grady of Anderson Design Associates

CLIENT
Gerry Baby Products Co.

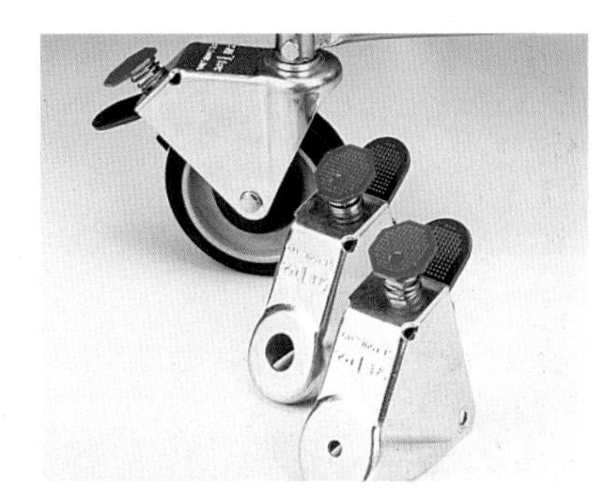

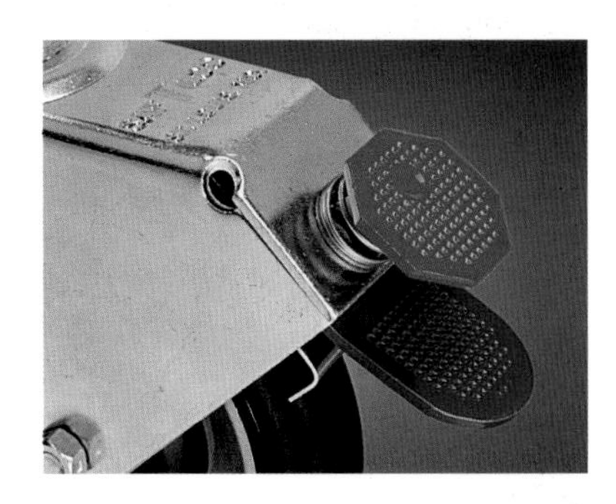

Saf-T-Loc

DESIGNERS
Bryce G. Rutter, Ph.D. of Metaphase Design Group;
Dan Houseman of Savvy Engineering

CLIENT
Saf-T-Loc

Pure Air Ultrasonic Humidifier

DESIGNERS
Glenn Klaus

DESIGN SCHOOL
Carnegie Mellon University

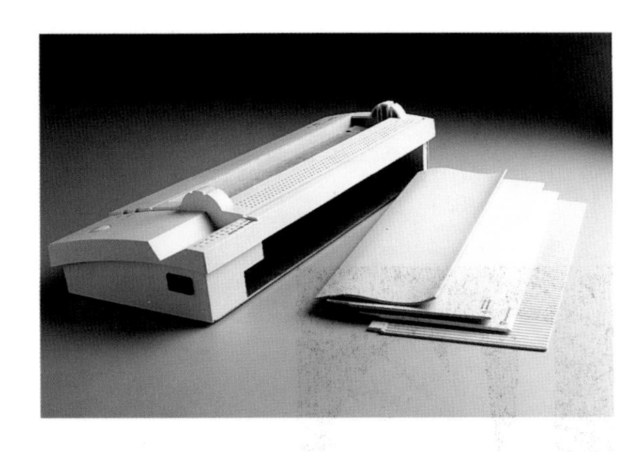

250T Therm-A-Bind®

DESIGNERS
Walter Herbst, James Caruso,
Thomas Milewski of
Herbst LaZar Bell Inc.;
Al Vercillo of General Binding Corp.

CLIENT
General Binding Corp.

Le Pet Cafe®

DESIGNER
Tim O'Donnell of
O'Donnell Pet Products

CLIENT
O'Donnell Pet Products

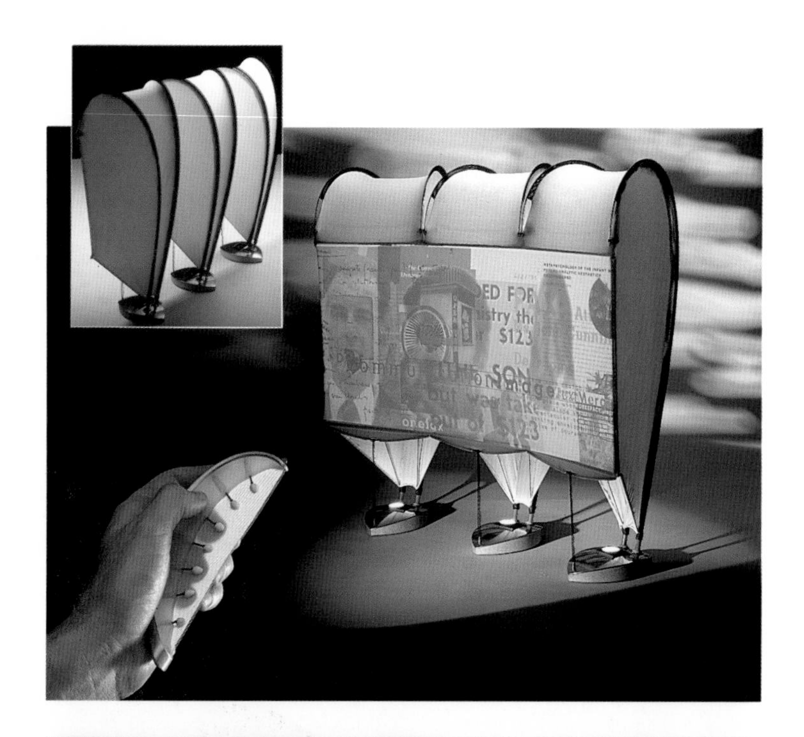

Nomadic Workstation

DESIGNERS
Donald Carr, Scott Makela

DESIGN SCHOOL
Cranbrook Academy of Art

SPONSORS
NCR Corp.; NYNEX

SmartLevel Series 200

DESIGNERS
Blake Wharton, Jay Wilson of GVO, Inc.; Edwin Seipp III, Andrew Butler of Wedge Innovations

CLIENT
Wedge Innovations

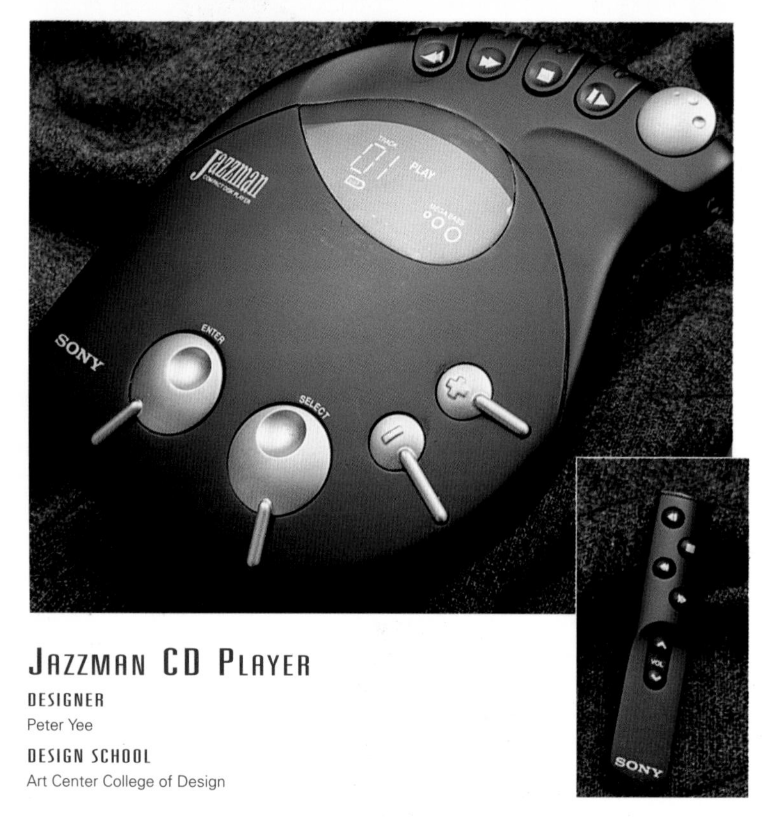

Jazzman CD Player

DESIGNER
Peter Yee

DESIGN SCHOOL
Art Center College of Design

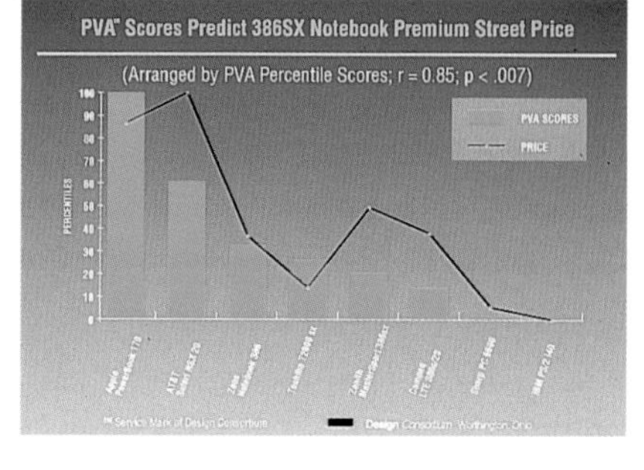

"Macintosh PowerBook vs. PCs" Product Value Analysis

DESIGNERS
Michael Barry of GVO; Ronald J. Sears of Design Consortium

CLIENT
Apple Computer, Inc.

Soft Notebook Computer

DESIGNERS
Emilio Ambasz, Eric Williams of Emilio Ambasz Design Group

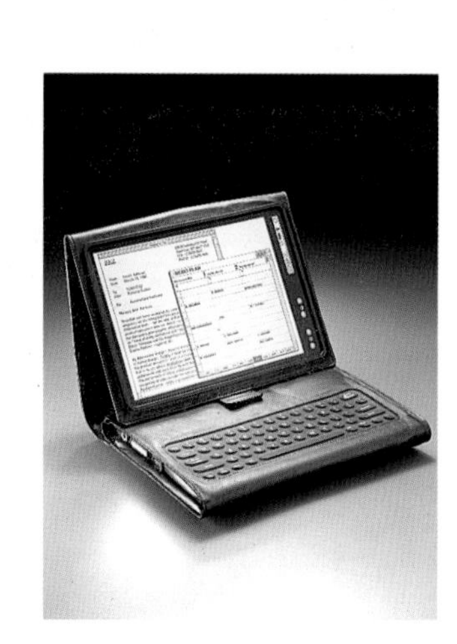

Cordless HandyMixer

DESIGNERS
Gary Van Deursen, Greg Hoffmann of Black & Decker Inc.

CLIENT
Black & Decker

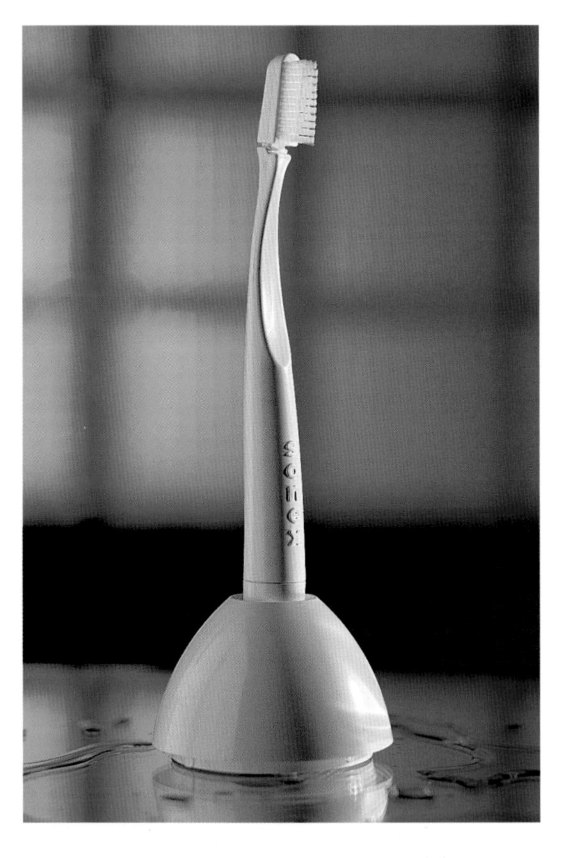

Ultrasonic Toothbrush

DESIGNERS
Benjamin Beck, John Costello of Design Continuum Inc.; Robert Bock of Sonex International Corp.

CLIENT
Sonex International Corp.

Storage Express

DESIGNERS
Sohrab Vossoughi, Robert Dillon, Christopher Alviar of ZIBA Design; John Archer, Ken Keeler of Intel Corp.

CLIENT
Intel Corp.

Scanjet IIc

DESIGNERS
Mo Khovaylo, James Dow of Hewlett-Packard

CLIENT
Hewlett-Packard

Family CPR Trainer

DESIGNERS
John Hamilton, John Cline, Charlie Patterson, Geof Garth of Laerdal California, Inc.; Einar Egelandsdal of Laerdal Medical

CLIENT
Laerdal California, Inc.

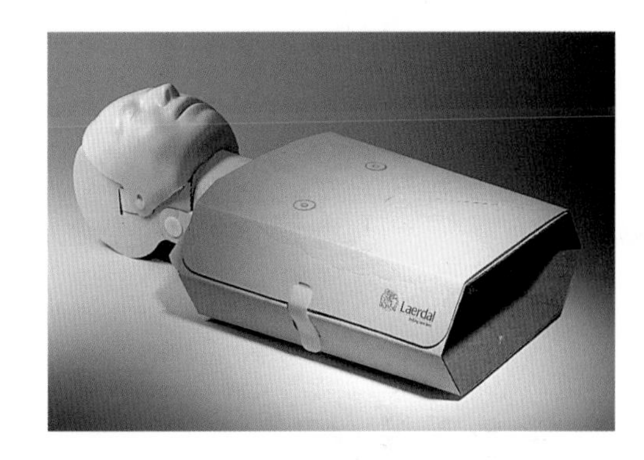

Cordless Mulching Mower

DESIGNERS
Alex Chunn, Adrian Hartz, Keith Long, Harvey Haynes of Ryobi Motor Products; Jim Watson, Doug Alsup of Alsup Watson Assoc., Inc.

CLIENT
Ryobi Motor Products

VCR VR5802/VR5602

DESIGNERS
John Costello, Michel Arney, Benjamin Beck of Design Continuum Inc.

CLIENT
Samsung Electronics Co., Ltd.

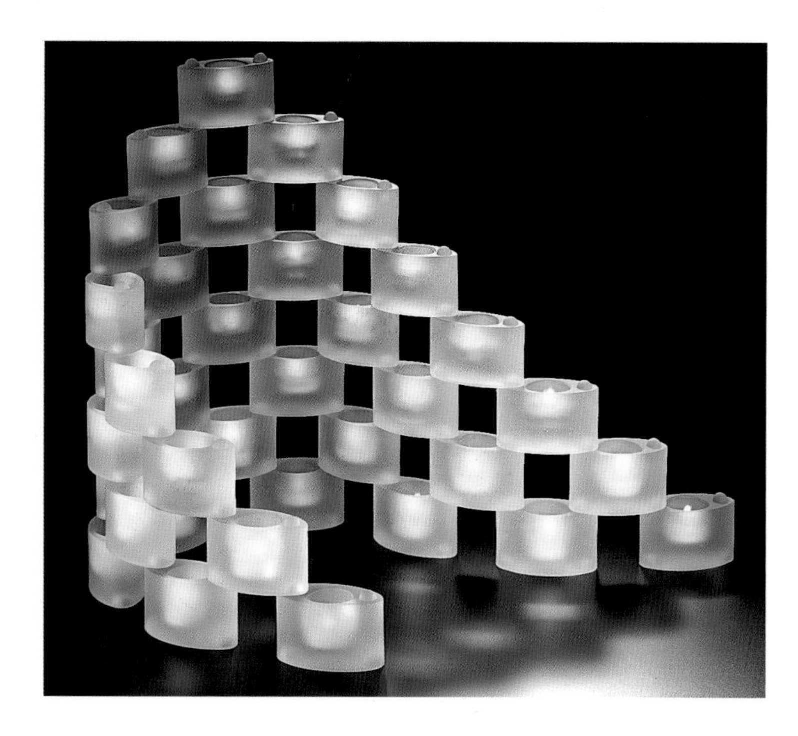

Catseye Candle Holder

DESIGNERS
Laura Handler, Dennis Decker,
David Peschel of Handler

CLIENT
Design Ideas

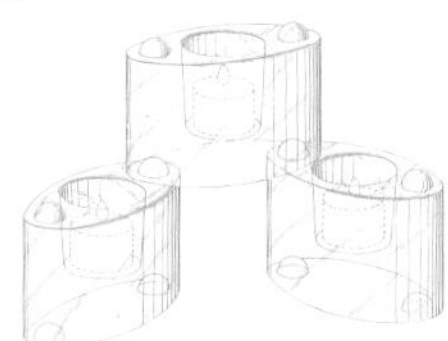

Backpacking Tents

DESIGNER
Charles Duvall of Moss Inc.

CLIENT
Moss Inc.

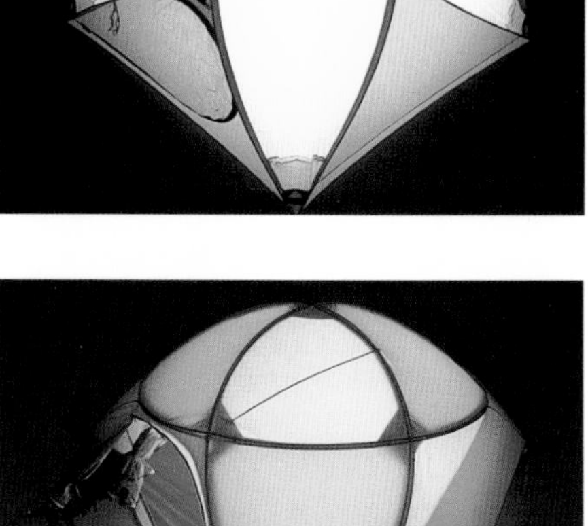

Monet

DESIGNERS
Daniel Klitsner, Mark Eastwood,
Derek Garvens of
Klitsner Industrial Design, Inc.

CLIENT
Memtek

Binoculars

DESIGNERS
Jorg Ratzlaff, Hanspeter Leins of
frogdesign inc.

CLIENT
Carl Zeiss

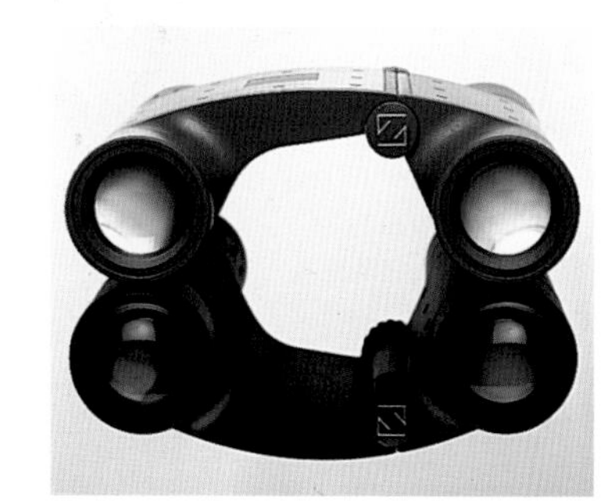

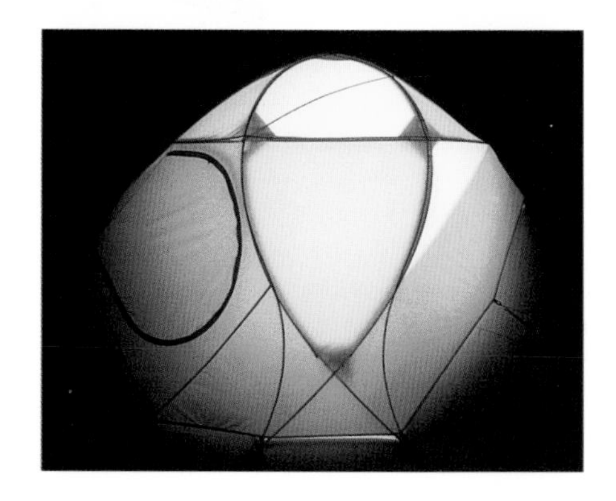

E2

DESIGNERS
Automotive Design Staff of
Designworks/USA

CLIENT
BMW Technik GmbH

Kool Mate 36

DESIGNERS
James J. Costello of
Igloo Products Corp.;
Udo Fritsch of
Data Electronics Devices

CLIENT
Igloo Products Corp.

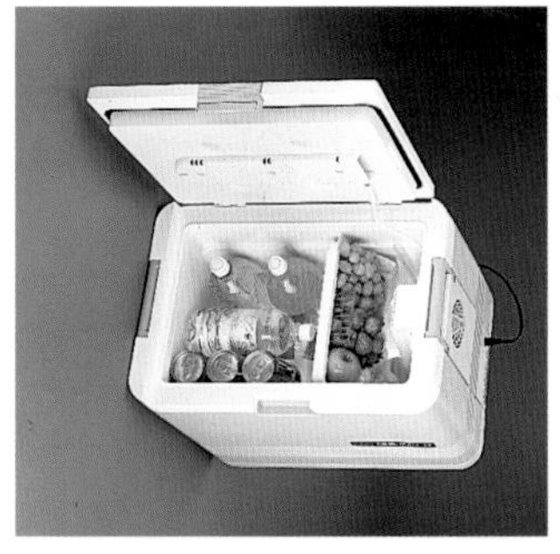

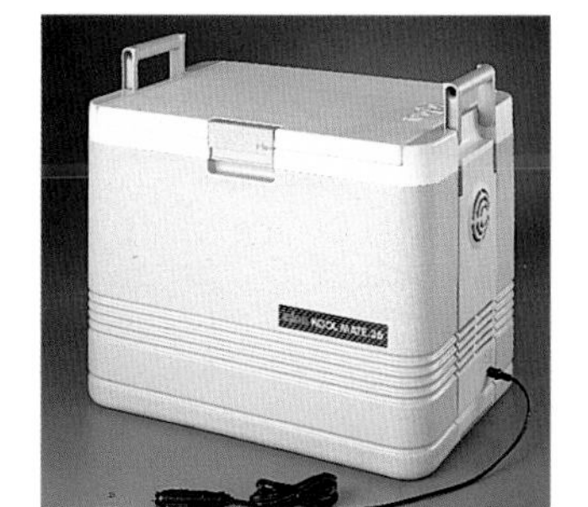

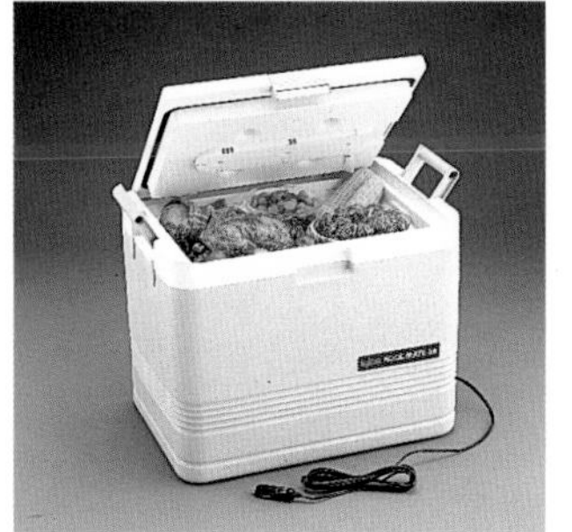

Activity Table

DESIGNER
Liz Krisel of Fisher-Price, Inc.

CLIENT
Fisher-Price, Inc.

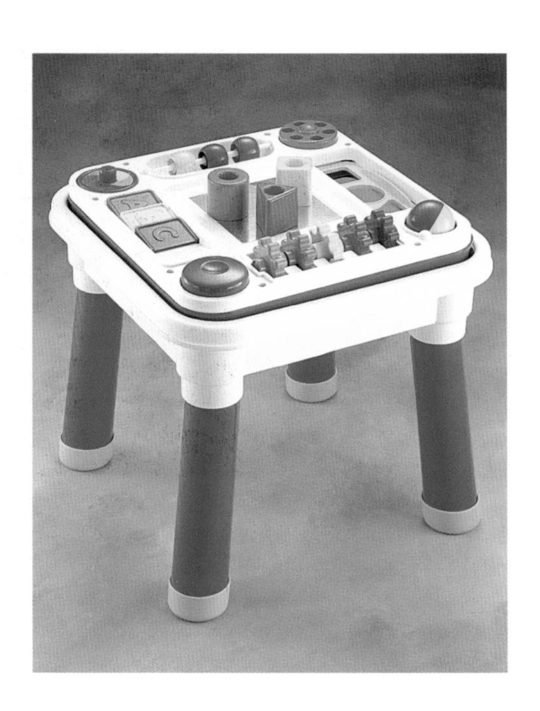

Fun Hydrant Sprinkler

DESIGNERS
Wade Maple, David Moomaw of
Fisher-Price Inc.

CLIENT
Fisher-Price, Inc.

Bulldog Chair

DESIGNERS
Dale Fahnstrom, Michael McCoy of Fahnstrom/McCoy Design

CLIENT
The Knoll Group

Playback

DESIGNER
Joe Ricchio of Ricchio Design

CLIENT
Atelier International Ltd.

Zwirl

DESIGNER
Benjamin Winter of Winter Design Mfg.

CLIENT
Zwirl Sales

Flash Tracks

DESIGNERS
The Fisher-Price Promotional Design Team

CLIENT
Fisher-Price, Inc.

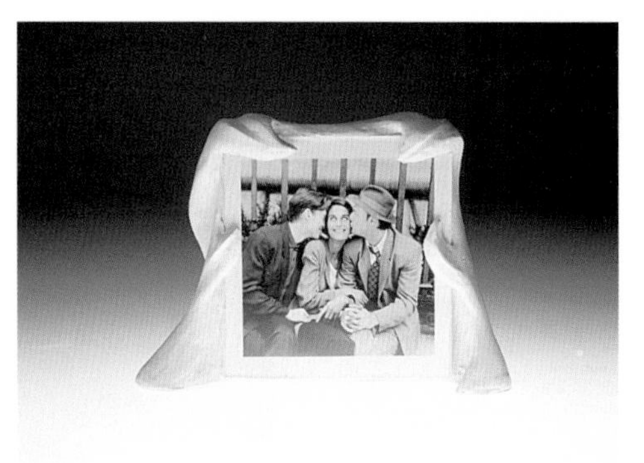

The Eternal Frame

DESIGNER
Mary Heather Worley

DESIGN SCHOOL
Pratt Institute

Multi-scale Fretboard

DESIGNER
Ralph Novak of Novax Handcrafted Guitars

CLIENT
Novax Handcrafted Guitars

Qualtronics Fax Machine

DESIGNERS
Bresslergroup Staff;
Qualtronics Staff

CLIENT
Qualtronics Corporation, Inc.

Morrison Network

DESIGNERS
Knoll Design Department

CLIENT
The Knoll Group

Computer Accessory Product Range

DESIGNERS
Tim Brown, Jane Fulton, Robin Sarre,
Bill George of IDEO Product Development;
Ken Krayer of Details

CLIENT
Details

Lincoln

DESIGNER
Peter Stathis

DESIGN SCHOOL
Cranbrook Academy of Art

BCN

DESIGNERS
Brayton International Collection Design Team

CLIENT
Brayton International Collection

Z-System

DESIGNER
the frogdesign team

CLIENT
Zenith Data Systems

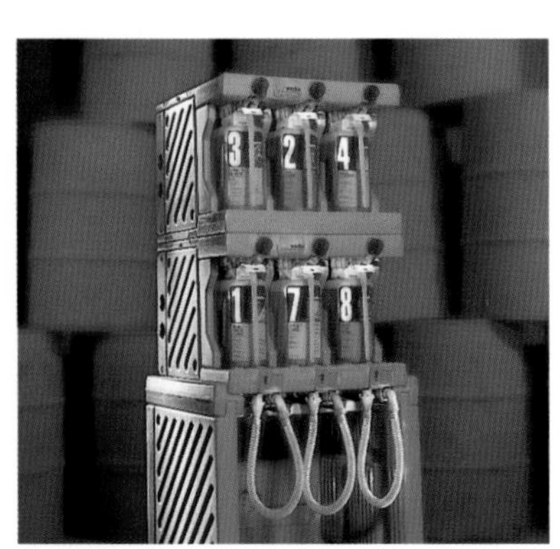

designing for Industry

WINNERS

LAISer II

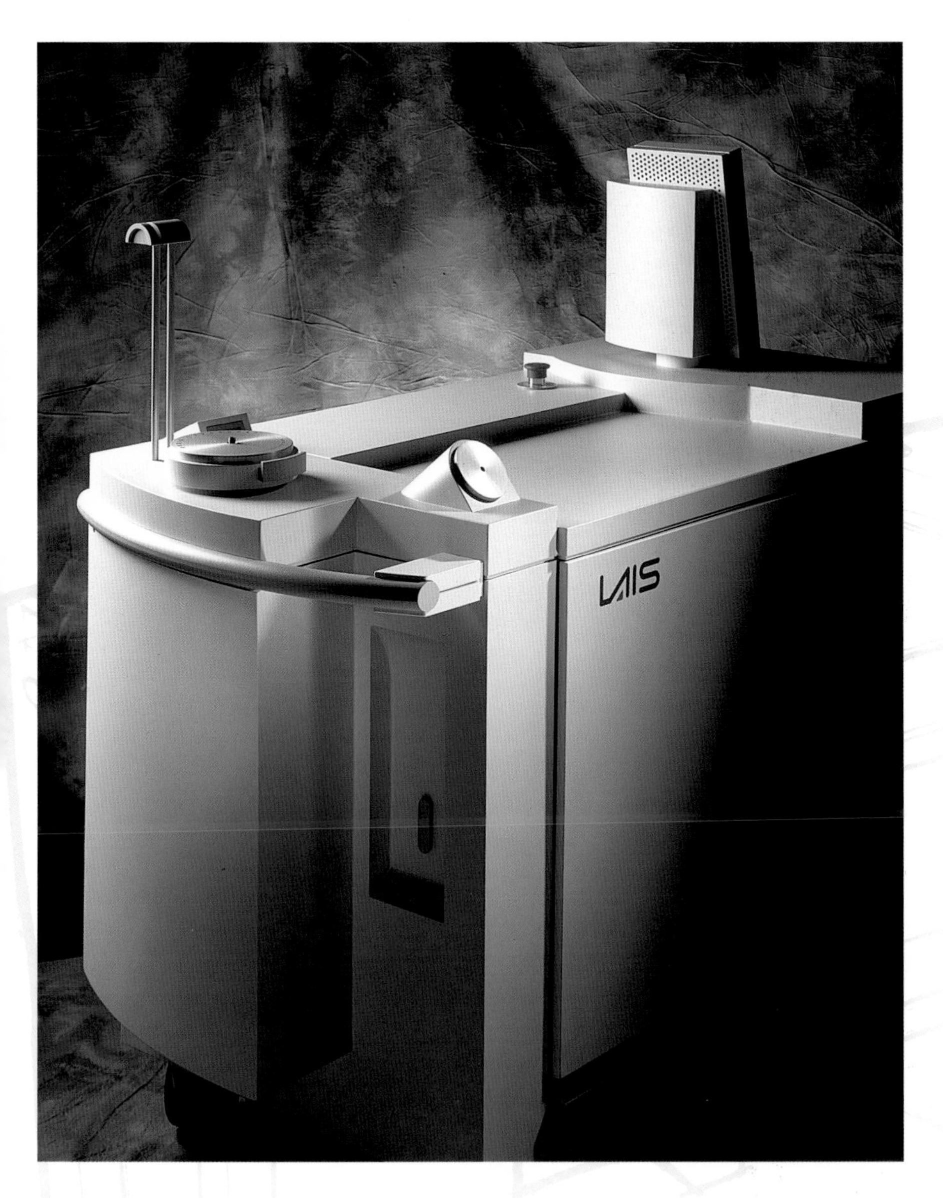

DESIGN OBJECTIVE

To design a second generation eximer laser that was adaptable and serviceable, put the user at the center of the design process, exceeded claims made by the competition, and provided a recognizable visual identity for the emerging technology.

DESIGN SOLUTIONS

The new arrangement of the internal components reduced overall height by more than a foot, and lowered the unit's center of gravity, making it more stable on ramps and assisting the structural integrity of the chassis. The noise level was reduced by putting rubber brushings on the clamps that harness the vessel to the chassis, and placing the cooling fan at the end farthest from doctor and patient.

RESULT

The new design of the LAISer II makes laser angioplasty much easier for cardiologists to use. The design reduced production costs by a factor of four.

OTHER AWARDS

–Industrie Forum Design Hannover, 1994
–1993 Minerva Awards

"...The designers have carefully enclosed the essential functioning of this high-technology equipment in gently curving forms that visually console the patient."

JUROR SHARYN THOMPSON, IDSA

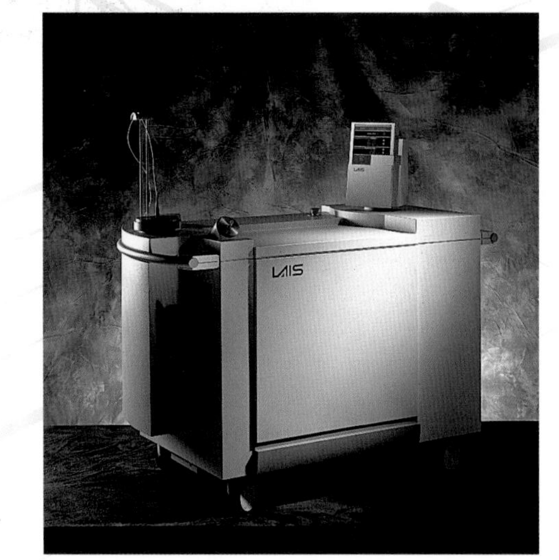

DESIGNERS

John von Buelow, Ernesto Quinteros, David Stocks, Karel Slovacek of SG Hauser Associates; Larry Reedy, Mark Whitebook of LAIS

CLIENT

LAIS Advanced Interventional Systems

Medical Ultrasound Imaging System

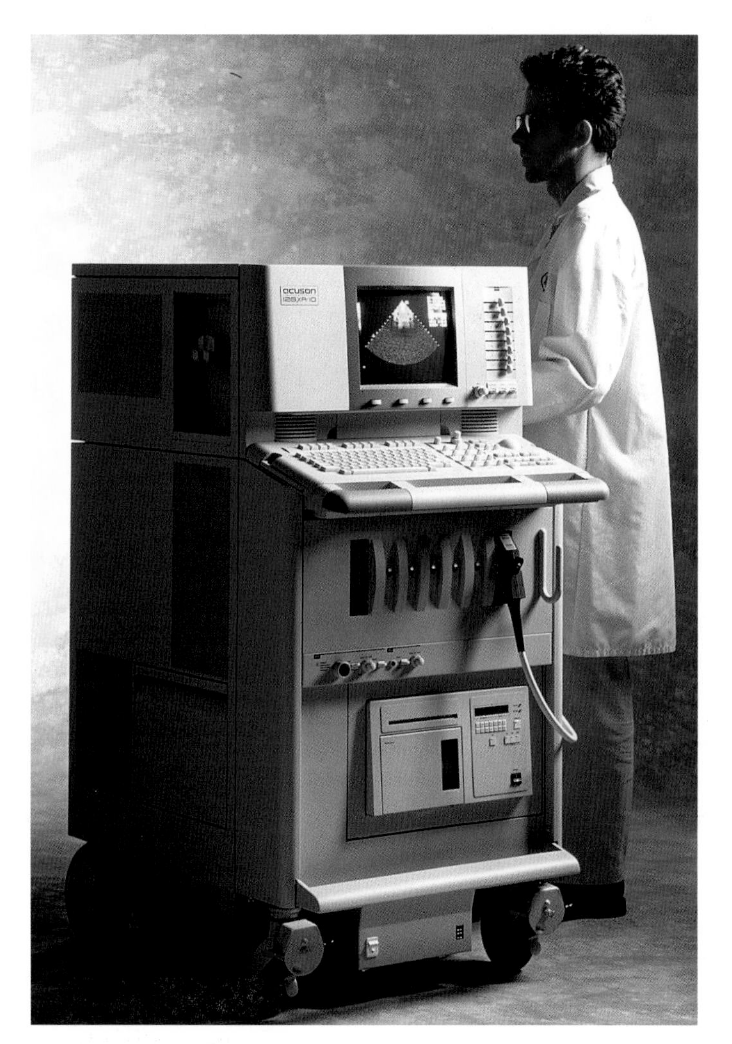

DESIGN OBJECTIVE

To design an ultrasound unit that would enhance user interaction and be suitable for the lower volume manufacturing processes used for this specialized machine while minimally impacting the unit's internal layout.

DESIGN SOLUTIONS

An emphasis on vertically oriented details and the large vertical corner radii makes the mass of the product appear less squat than the earlier design. The corner radii, the consistent color for keys and body, and an emphasis on the integration of several elements of the product present a friendlier, less intimidating feel to the user. The rubber handle details double as corner bumpers while providing an extra degree of comfort for the operator.

RESULT

Specific improvements include: better system lighting; easily accessed and convenient probe storage; a footrest that helps manage probe cables; and redirected airflow for noise reduction. Acuson stock went up as a direct result of the product's introduction.

OTHER AWARDS

–*I.D.* Annual Design Review

"This ultrasound system's design is innovative, with exquisite attention to detail. Well-organized, its control panel design is most successful, with a clear delineation of functions...."

JUROR JERRY HIRSHBERG

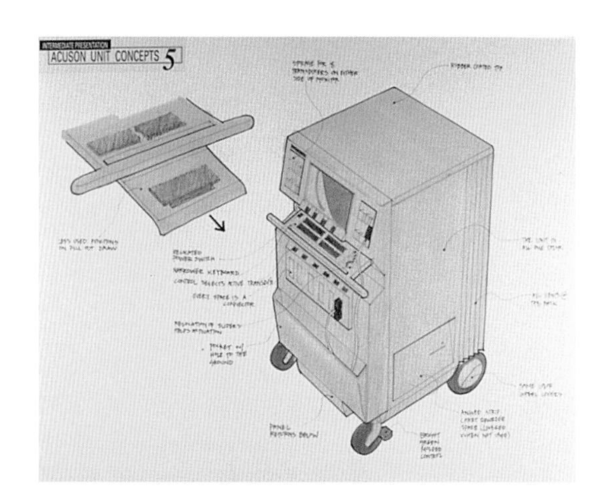

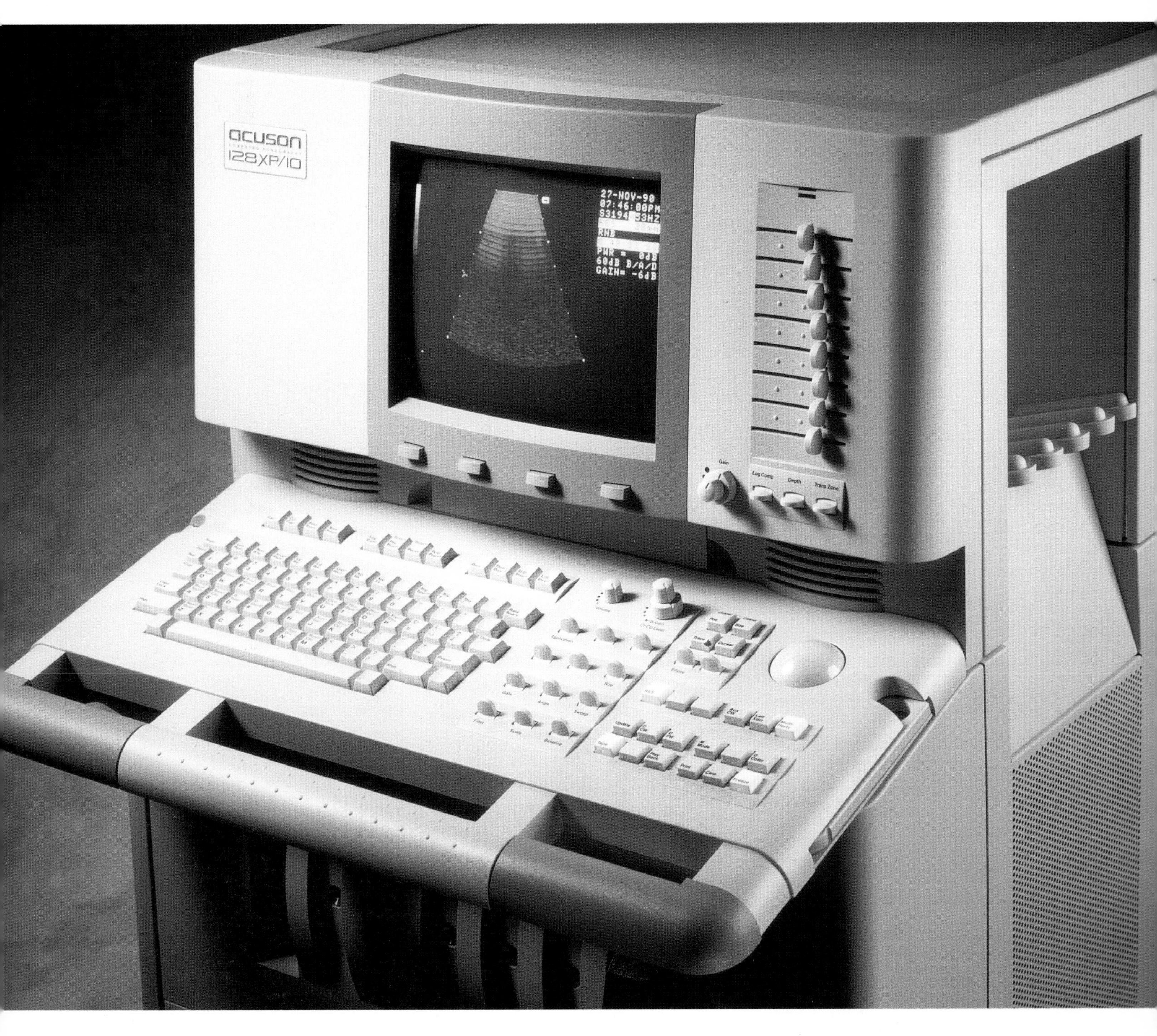

DESIGNERS
Tim Parsey,
Naoto Fukasawa of ID TWO;
Rich Henderson of
Acuson Corp.

CLIENT
Acuson Corp.

7785 Service Bay System

"IBM's diagnostic computer is a rare combination of technological sophistication and industrial good looks. Performance and aesthetics mesh perfectly...."

BUSINESS WEEK, MAY 24, 1991

"...With bumpers that bump, handles that say 'grab me' and wide-stance wheels that can negotiate grates, hoses and cables like a moon rover, this product delivers in performance the reliability, endurance, stability and technological sophistication that its visual semantics promise...."

JUROR ARNOLD WASSERMAN, IDSA

DESIGN OBJECTIVE

To develop a total workstation for the service technician that would be used to diagnose and repair, update technicians on recalls, send and receive technical notes, and order parts. The computer hardware had to be integrated into the unforgiving environment of a service garage.

DESIGN SOLUTIONS

Comprised of four primary components, the system weighs approximately 350 pounds, and is 63" x 27" x 31". The main body houses the CPU, including both optical and storage devices. The platform for the 19-inch display serves as a protective garage in which the keyboard is stored, while the wedged surface discourages users from placing beverages on it which could spill and damage the system. Oversized bumpers protect both the system and the automobiles, and covers protect the disk drives.

RESULT

The Service Bay System is a sophisticated, state-of-the-art computer that links to the automobile's on-board computer as well as to local area networks and satellite networks.

OTHER AWARDS

–*I.D.*, July/August '91, Honorable Mention
–Chrysler's 1991 Pentastar Award

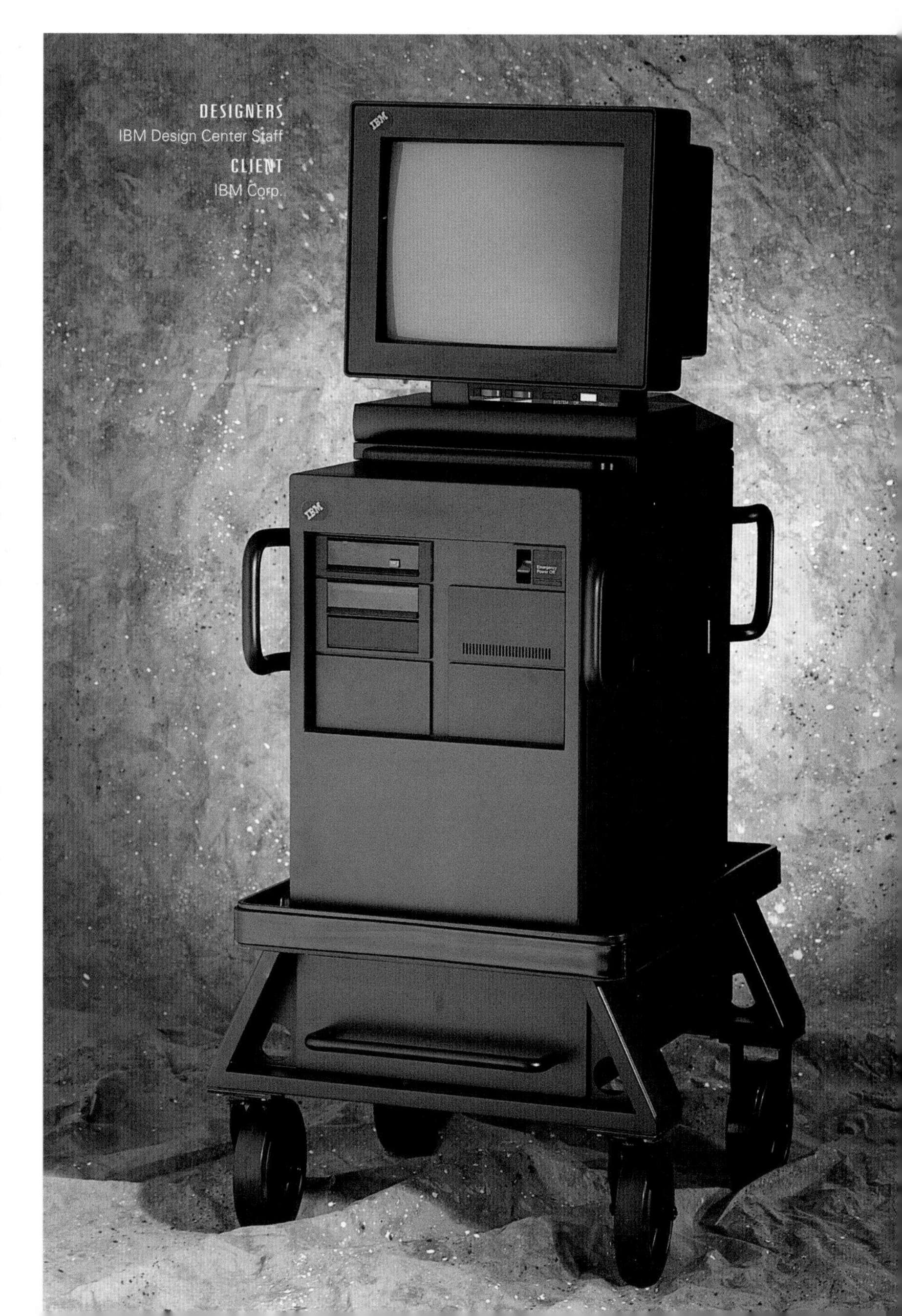

DESIGNERS
IBM Design Center Staff

CLIENT
IBM Corp.

Relay Furniture

DESIGNERS
Geoff Hollington of Hollington Associates;
Susan Monroe, Don Goeman of Herman Miller, Inc.

CLIENT
Herman Miller, Inc.

"Unlike traditional systems of furniture, the elements of Relay...do not lock together in a rigid fashion. They come together casually, simply, butting up against one another. They don't lock, they dock."

I.D., SEPTEMBER/OCTOBER 1990

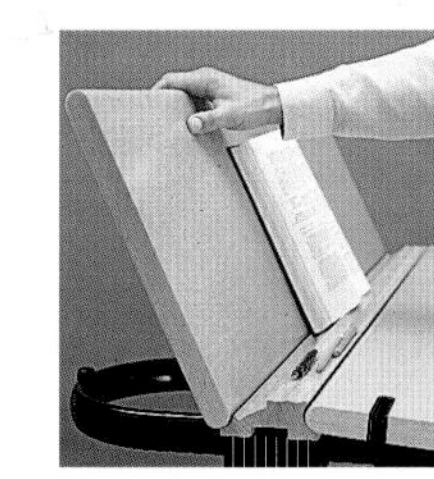

DESIGN OBJECTIVE

To create a system of freestanding user-friendly furniture which would anticipate the trend away from conventional component systems as technology becomes more enabling yet physically softer.

DESIGN SOLUTIONS

The product's most obvious innovation is its mobility and, thus, flexibility. Because adjacent furniture pieces never physically connect and all pieces move easily across the floor, users can reconfigure their work environments themselves.

RESULT

The Relay system caters to the needs of its users in very basic ways, without being dated by the average 3-year life span of a piece of equipment. Accessories address the needs of the current generation of technology.

OTHER AWARDS

–*IBD and Contract Magazine*,
1991 Product Design Award
–Industrie Forum Design Hannover,
iF Award for Good Industrial Design

Anthropometric Measuring System—"Anthropometron"

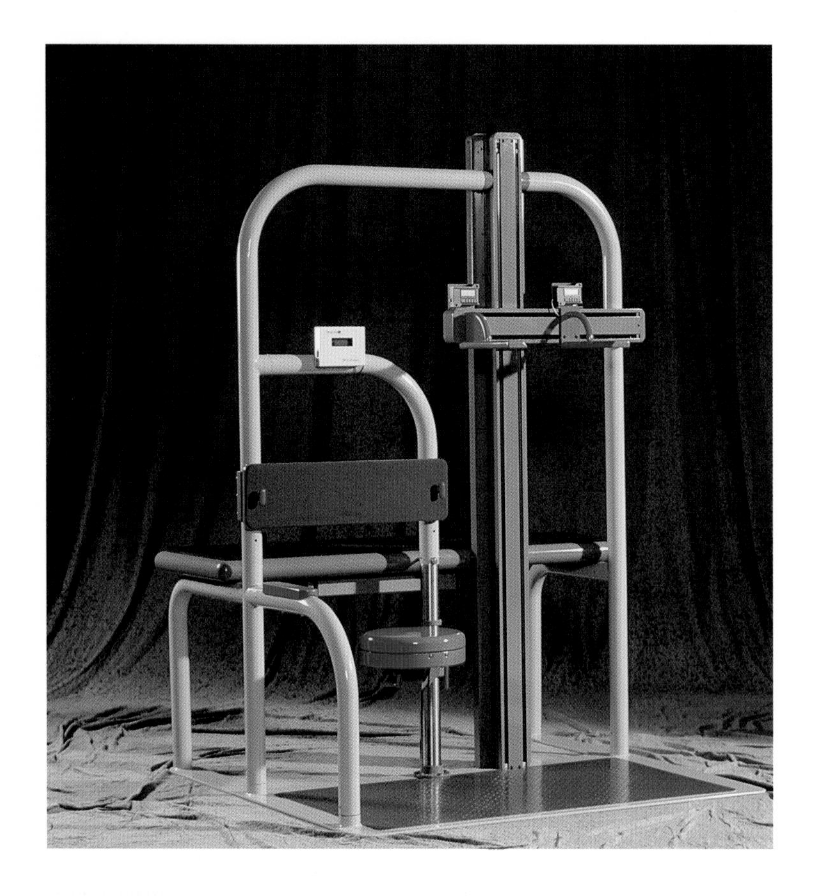

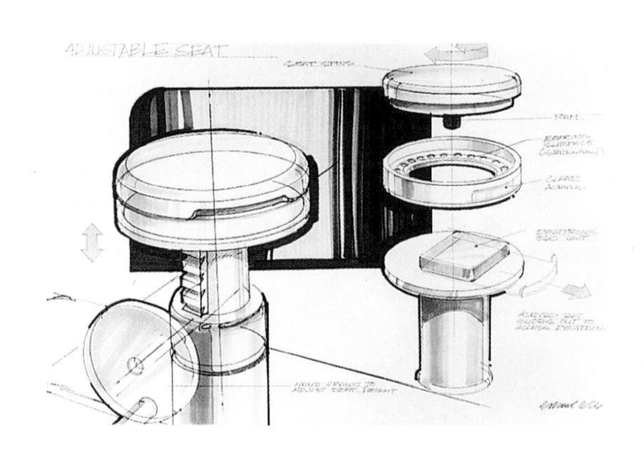

"The AMS makes the science of ergonomic measurement palatable for both the subject and the scientist. It brings the anonymous art of gathering human dimensions out of the laboratory closet and makes it friendly and nonthreatening, functional and efficient...."

JUROR CHARLES ALLEN, IDSA

DESIGN OBJECTIVE

To develop a human body anthropometric measuring system that would rapidly collect body size and range of motion data from individual subjects. The device had to be easily dissembled, measure linear dimensions accurate to 1 mm and angular measurements to the nearest degree, and be able to fit through a door.

DESIGN SOLUTIONS

A biaxial system eased the operators' task of obtaining many distinct measurements through a horizontally sliding caliper mounted on a vertically sliding beam. Sliding along a bearing track, electronic calipers obtain point-to-point measurements, which are then displayed on an LCD. An OEM infrared emitter with an LED display locates the center of pivotal rotation and displays the amount of rotation in degrees.

RESULT

The one-of-a-kind device reduces operational time for the entire series of measurements from 90 minutes per subject to under 30 minutes, and can measure two subjects at once.

DESIGNERS

Steven W. Ward, Mark A. McLean, Keith D. Shapland of Walter Dorwin Teague Assoc.; R. Conway Underwood, Rush Green of Boeing PHI Group

CLIENT

Boeing Commercial Airplane Group

IMAGERING

DESIGN OBJECTIVE

To solve the deterrent influence of existing mammography equipment by designing a device that responds to patient needs.

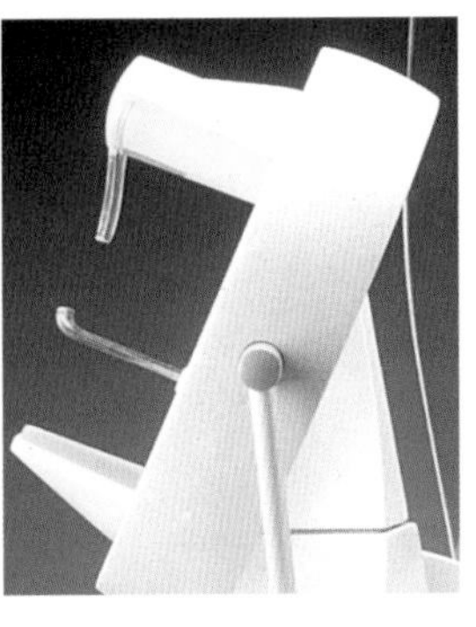

DESIGN SOLUTIONS

A ring-shaped gantry allows the patient to face the technician instead of the machine, while lessening the pain of the examination since the patient spends less time with the breast compressed. Colors from the cool end of the pastel spectrum replace the conventional, often condescending, pink pastels. The molded plastic construction and a heating element placed in the compression plates address frequent complaints that the machines are cold to the touch.

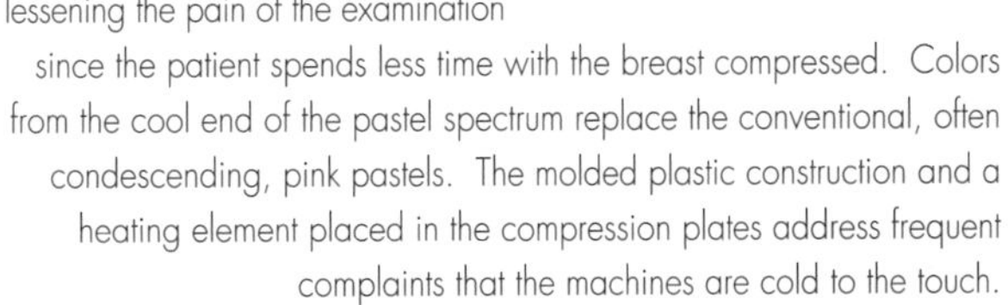

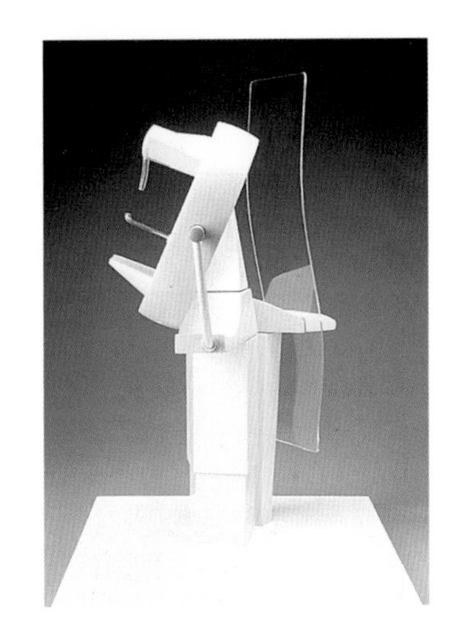

RESULT

To appeal to the doctors and hospitals purchasing the equipment, ImageRING reduces the time required with each patient, increasing profits.

OTHER AWARDS

–The Sixth Symposium on Healthcare Design, 1993 Healthcare Design Competition Award

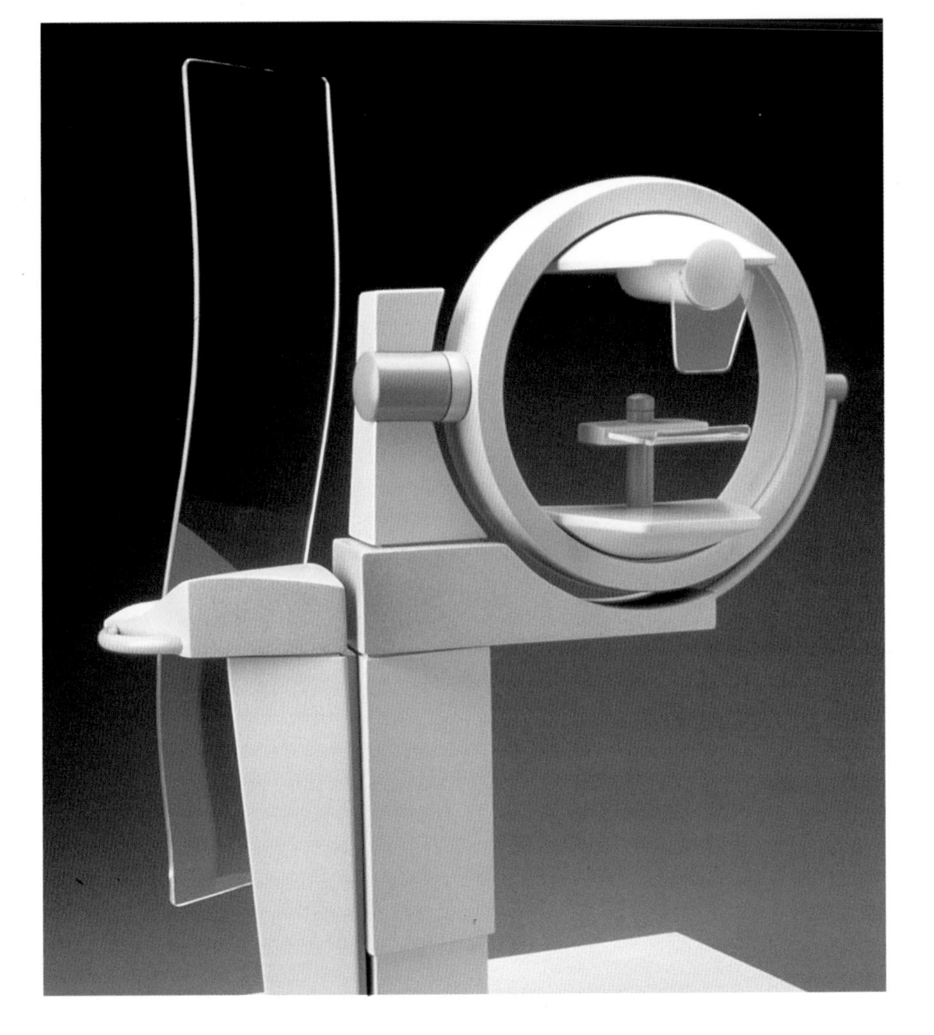

DESIGNER
Jane Saks-Cohan

FACULTY ADVISORS
Mark Lacko, Joseph Parriott

DESIGN SCHOOL
Pratt Institute

"The imageRING dramatically improves on the design of existing mammography equipment. Its design shows a genuine sensitivity to the human encounter with this product and process...."

JUROR VINCE FOOTE, FIDSA

Large Screen LCD Video Projector

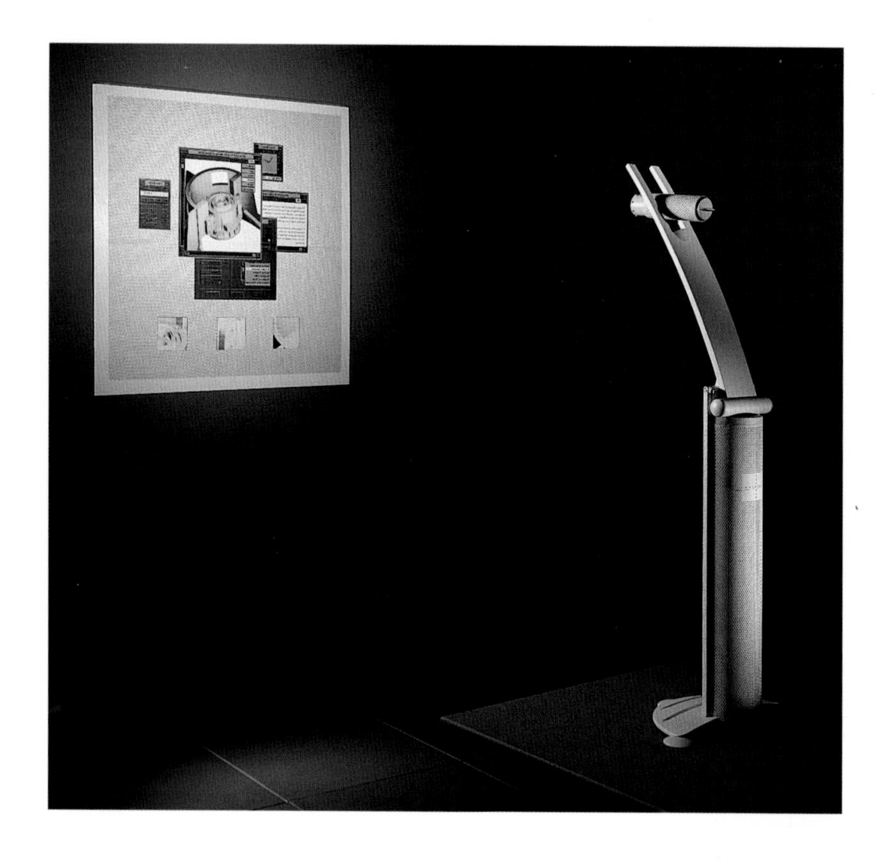

DESIGN OBJECTIVE

To design a portable LCD Video Projector with a large flat screen, small footprint, and enough flexibility for viewing by one person or a large group.

DESIGN SOLUTIONS

The LCD Video Projector can be used on a desktop and in a small conference room, since the image and screen size are adjustable with an auto focus system accommodating different viewing needs. The still images or video can be controlled by a keyboard or a remote device, and a speaker in the rear cylinder provides audio capabilities.

RESULT

The large screen LCD video projector, with audio input and output capabilities, has a desktop footprint 88 percent smaller than standard color CRTs. The articulated arm allows the image to be projected onto an 18-inch screen for desktop use or onto a 36-inch screen for group presentations. The projector is lightweight and compact. The screen folds into quarters and the angle of the desktop screen can be set at 5 degrees or 10 degrees to accommodate a range of viewing heights.

"Successful design is design that makes peoples' lives easier in some way, solves at least one problem, and delights and engages the eye and the soul!"

DIGITAL EQUIPMENT CORP.

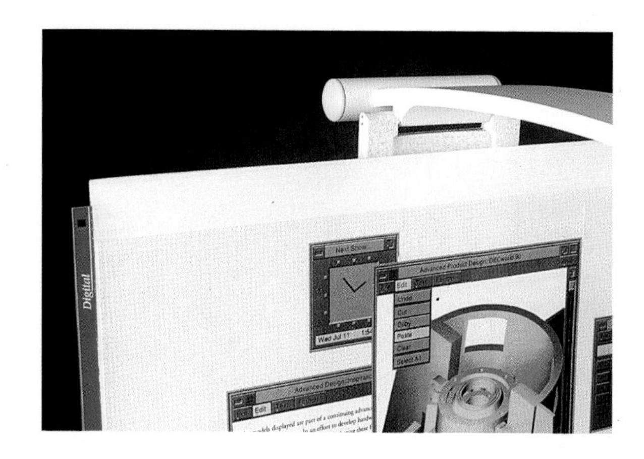

"...This design is exciting because of its complexity—the multiple configurations of the device and layering of detail. The unusual exploitation of cliché materials and forms explores a new business machine language for products we want to relate to emotionally."

JUROR TUCKER VIEMEISTER, IDSA

DESIGNERS
Meg Hetfield of Digital Equipment Corp.

CLIENT
Digital Equipment Corp.

Edit
Undo
Cut
Copy
Paste
Clear
Select All
Autumn Leaves
Spring Flowers
Summer Hills
Shades of Blue
Charcoal & Pinstripe

LEAPFROG

"...We were fascinated to see this three-dimensional shape represented on an exceptionally thin two-dimensional surface...."

JUROR SHARYN THOMPSON, IDSA

DESIGN OBJECTIVE

To make computing feel natural by embodying emerging natural interfaces and advanced technologies, and then organizing them to adapt to the work patterns, environment and unique ergonomic needs of office professionals.

DESIGN SOLUTIONS

In addition to optimizing the system for pen and voice input, research showed the need for a keyboard with an integrated pointing device. A digitizer, processor, batteries and other microelectronics would need to be fully integrated to eliminate restrictive cables. To overcome the unnatural thickness of the tablet, the edges were beveled in two directions, creating a flush-to-the-desk writing surface in both portrait and landscape orientations. The glass surface is flush to the top and textured, so that when you use a stylus on it, it feels like a pencil on paper.

RESULT

The working prototypes yield realistic research data that influences current and future product development at IBM.

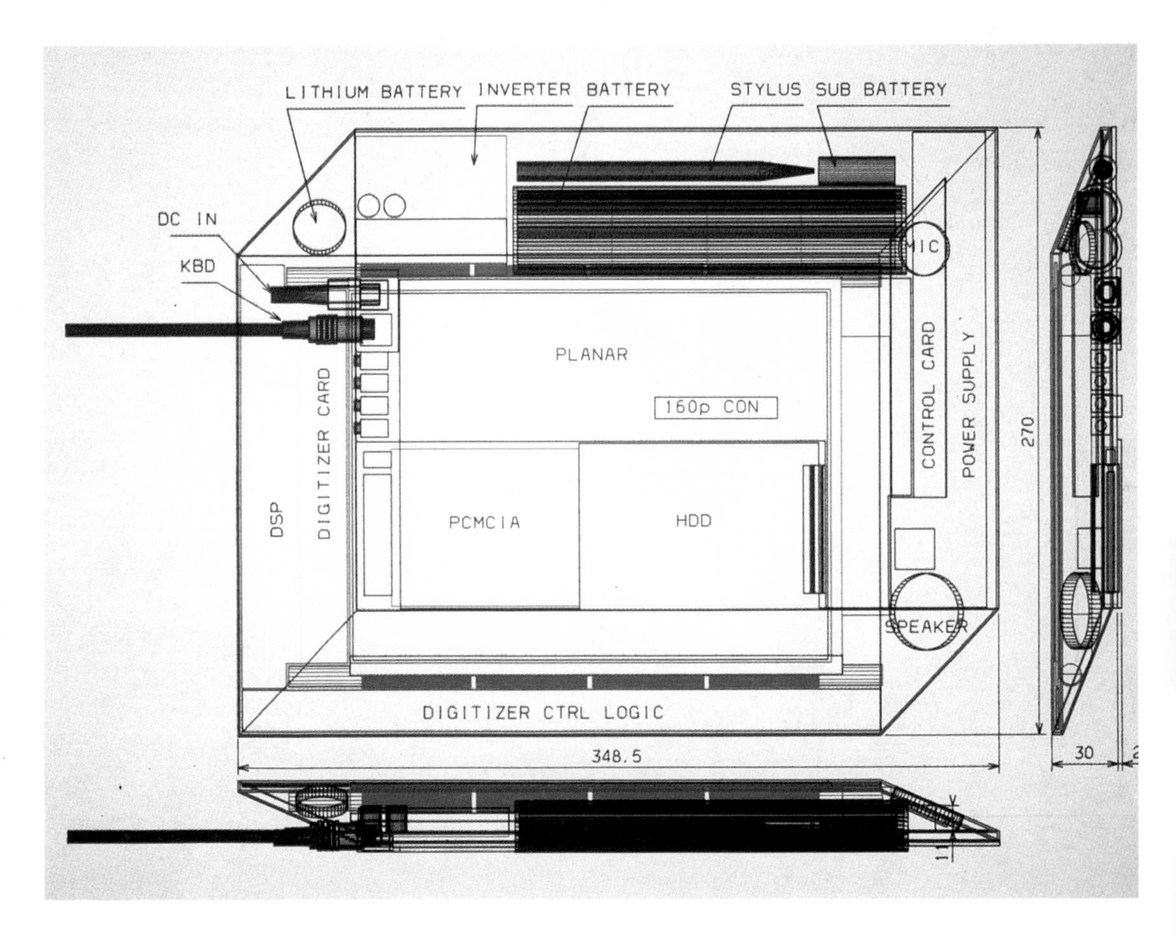

DESIGNERS
Richard Sapper;
Samuel Lucente of IBM Corp.

CLIENT
IBM Corp.

3495 Tape Library DataServer

DESIGNERS
Martin J. Marotti,
James L. Lentz,
Bert Slawson of IBM Corp.

DESIGN OBJECTIVE
To create a data library of immense capability with the following requirements: reliability, expandability, ease of installation, comprehensive data management, and investment protection.

DESIGN SOLUTIONS
The 3495 DataServer provides a fully automated tape cartridge management system for customers who have traditionally sorted by hand through thousands of cartridges to find specific data, such as customer account information. With the 3495, a remote operator simply punches an account number in, and a robot retrieves the proper cartridge, inserting it into a tape drive. The 3495 is modular in size and capacity and installs quickly. All controls can be reached while seated, making it accessible to the disabled. Moreover, the designers developed a detailed end-of-life disposal program for recycling.

RESULT
Customers are thrilled with the way this product fully addresses all their requirements, and they appreciate the fact that the technology was made fun and exciting.

OTHER AWARDS
–Industrie Design Hannover,
1993 iF Award for Good Industrial Design

"...Humans may not need to walk inside this thing often, yet this design humanizes the hard edge of super-sophisticated technology."

JUROR CHIPP WALTERS, IDSA

LiveBoard

DESIGN OBJECTIVE

Observations about business practices combined with research into anthropological and sociological aspects of work practices demonstrated the need for tools that would facilitate shared spaces and group work surfaces.

DESIGN SOLUTIONS

The internal hardware and controlling optics are mounted on a sliding rack for ease of assembly, alignment and maintenance. Human factors data and full size mock-ups indicated that the optimum work surface for the physical reach range and line-of-sight visibility was 67 inches on the diagonal. CAD-generated optical paths facilitated the iterative process of locating the interior components while maintaining a minimum exterior envelope.

RESULT

The LiveBoard combines the simplicity of a whiteboard with the power of multimedia. Users interact with it through an ergonomically designed, wireless stylus which can be used directly on the board or as a pointer/control device from a distance. The initial production orders were completely sold out, with back orders taken.

OTHER AWARDS

–1993 Good Design Award

"...It's almost magic. The simple, intuitive interaction capability and provision for multiple users enhances spontaneity, which will spark constructive dialogue and better business solutions."

JUROR CHARLES ALLEN, IDSA

"Successful design is embodied in an object that has the needs of the user, the manufacturer, and the client, balanced with the needs of the environment. This balance includes outstanding appearance and function at an appropriate cost, and a product life cycle that provides for renewal and recycling not simply death and burial."

PALO ALTO DESIGN GROUP

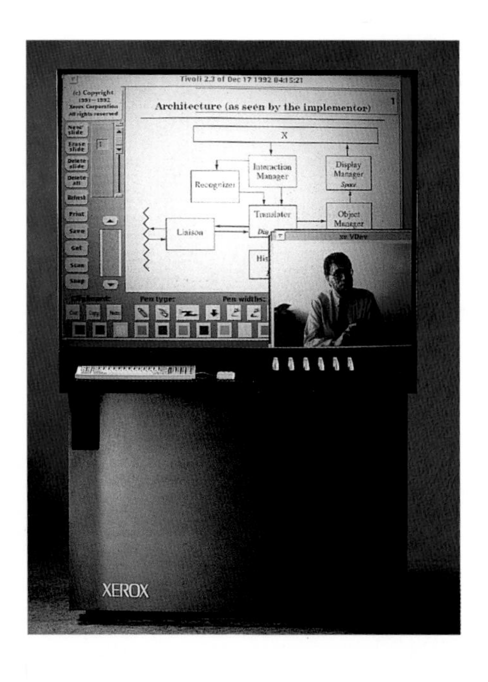

DESIGNERS

Malcolm Smith, Dallas Grove of Palo Alto Design Group;
Xerox Palo Alto Research Center (PARC)

CLIENT

Xerox

Electrical Vehicle Charging System

DESIGNERS
Mark Biasotti, John Lai,
Sigi Moeslinger, Mark Roemer of
IDEO Product Development;
Dick Bowman of Hughes Aircraft

CLIENT
General Motors/Hughes Aircraft

DESIGN OBJECTIVE

To create a system which would employ an off-board inductive charging system at the high power level needed to efficiently charge electric cars for day-to-day use.

DESIGN SOLUTIONS

The various charging schemes could be classified by two basic requirements—where the EV could be charged, and the time it takes to charge for an acceptable range. The IDEO and Hughes study found that four different chargers would be needed to meet the level of convenience required for an EV operator: a wall-mounted residential module; a curb-side column for use in cities at parking garages and at homes without garages; a kiosk-style energy system which would charge an EV in 15 minutes, located at convenience stores, parking garages, gas stations, etc.; and finally a portable adapter unit that would provide the ability to charge an EV at any 110 VAC outlet.

RESULT

This design proposes a family of devices for charging electric vehicles, using the familiar forms of gasoline fueling; however, rather than metaphoric gimmicks, these visual references act as inviting semantic bridges which in no way compromise functional safety, ergonomic performance and convenience.

"I have never encountered a finer demonstration of how advanced design can give marketable form to emergent technology, nor a better argument for just what industrial designers bring to new product innovation that engineers and marketers alone cannot...."

JURY CHAIR
ARNOLD WASSERMAN, IDSA

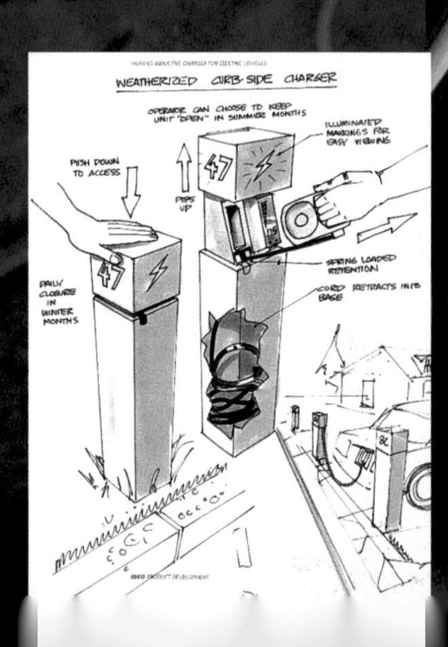

Ultralite

"...It is an extremely elegant solution to a complex design problem: how to achieve minimum aerodynamic drag and vehicle weight and maximum interior package with a design statement that is both rationally and visually appealing...."

JUROR FRITZ MAYHEW, IDSA

DESIGNERS

Charles M. Jordan, Jerry P. Palmer, Jim Bieck of General Motors Design

CLIENT

General Motors Corp.

DESIGN OBJECTIVE

To create an aerodynamically efficient vehicle that would get 100 mpg by halving the weight and doubling the fuel efficiency of a contemporary mid-size sedan.

DESIGN SOLUTIONS

The shape of the Ultralite evolved into an organic, tapered teardrop shape with the front track 5 inches wider than the rear track. The Ultralite body is constructed of carbon fiber, a material widely recognized for its strength-to-weight capabilities. Developed out of a duoflex material, the seats are lightweight, and provide superior ventilation and increased comfort. The rear-engined power pod module can be detached from the car as a unit, allowing different powertrains (such as electric or turbine) to be used.

RESULT

Achieving 100 mpg on the highway at 50 mph, the car's small frontal area and tapered shape make possible a drag coefficient of 0.192. Although it has the same amount of interior space as a Chevy Corsica, it is 18 inches shorter and 4 inches narrower. GM research scientists are currently working on a breakthrough carbon fiber manufacturing method called Pyrograph that would make mass production of the Ultralite cost effective, and therefore plausible.

AtLite Exit Sign

"...Its design improves legibility and actually enhances the interior environment. Any architect or interior designer will pick this product for their project because of the aesthetics alone, but the energy and labor issues make its selection compelling."

JUROR NOEL MAYO, IDSA

DESIGN OBJECTIVE

To create a thin, unobtrusive, and visible exit sign without frames or bezels that would require a minimal cut out to install.

DESIGN SOLUTIONS

To mold the panels absolutely flat and with no distortions or aberrations, the molder "shot" the panels while the tool was in progress. Internal reflection within the plastic panel maximizes the monochromatic light output of the LEDs. The selection of the *E-X-I-T* letter screen inks controled the LED and panel internal optics.

RESULT

Installation requires a much smaller opening in ceiling surfaces, less than half the opening size of existing products. Moreover, the edge-lit design gains an 80 percent energy savings over other products and never requires relamping.

DESIGNERS

Will Goldschmidt, Chris Carmody, Irving Schaffer of Designspring, Inc.; Robert Feldstein of Scientific Prototype; Harry Herz of Autronic Plastics

CLIENT

AtLite Lighting Equipment, Inc.

FirstPlay

DESIGNERS

Kevin Owens of PlayWorld Systems/PlayDesigns

"FirstPlay is an uncommonly inviting and joyous execution of the playground/fort. It beckons the child to touch and learn in this tiny world, in absolute safety."

JUROR RALPH OSTERHOUT, IDSA

DESIGN OBJECTIVE

To create a play structure made of soft materials that is fully accessible to all children, easily sanitized, usable indoors or outdoors, affordable, and easily assembled by one or two day care workers.

DESIGN SOLUTIONS

Adapting the turn-to-adjust/fold-over-to-lock approach of quick release bicycle wheels, the Camlocks are covered with childproof closures and provide strong, durable connectors without the need for tools. The product consists of three primary materials—galvanized, powder-coated steel, shock-absorbing, brightly colored, weatherproof cast foam polyurethane and a tough, fire resistant awning fabric.

RESULT

FirstPlay presents fun skill development for children 8 to 36 months old. FirstPlay is opening up new markets for the company in children's hospitals, doctors' waiting rooms, and resort hotels, among others.

OTHER AWARDS

–*I.D.* 1993 Annual Design Awards, Gold Medal
–Parent's Choice Award 1993

670 Perimeter

DESIGNERS
Bryan Hotaling,
Jon Rossman of Product Insight, Inc.;
Patrick Thornton of Thornton Design, Inc.

CLIENT
Tomey Technology, Inc.

"...The 670 beckons a patient to the dome with spots of light that will inform the clinician during a reduced test cycle. The soft, soothing forms reduce neck strain, resulting in better test accuracy and patient comfort...."

JUROR SHARYN THOMPSON, IDSA

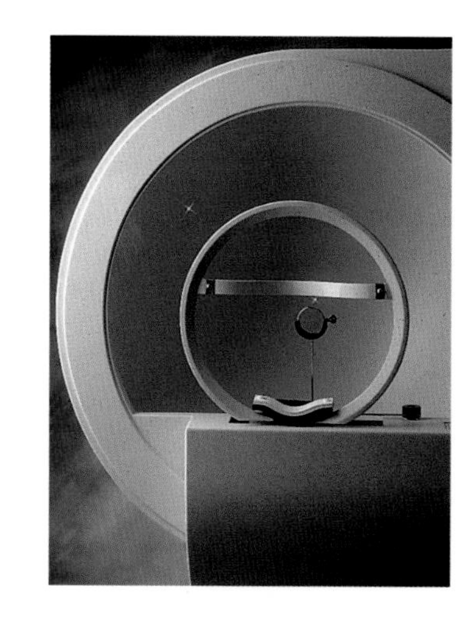

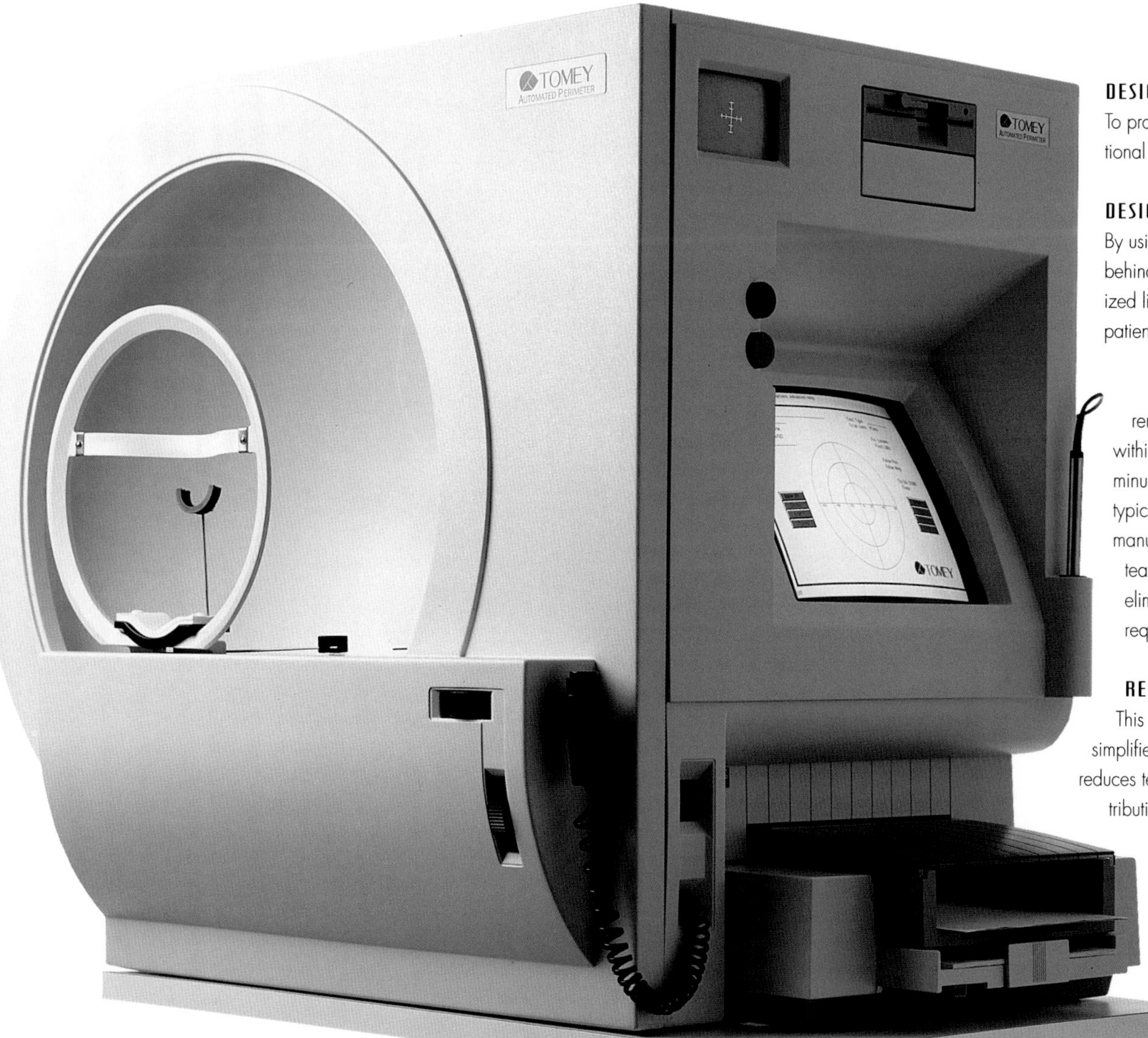

DESIGN OBJECTIVE

To produce a more ergonomic and functional product while selling it for less.

DESIGN SOLUTIONS

By using an array of 300 LEDs mounted behind the dome rather than a motorized light source projecting on the patient side of the dome, the front surface of the machine was pushed back considerably, removing the patient's head from within the device. A single cast aluminum head support ring replaced the typical "goal post" assembly. From a manufacturing standpoint, the design team saved on component costs and eliminated the calibration time required for a motorized system.

RESULT

This design improves patient comfort, simplifies instrument operation, and reduces test cycles. With an extensive distribution network in Europe and Japan already established and an expanded local R & D and manufacturing facility in place, Tomey is now ready to attack the market leader in the US.

Suburban Public Phone

DESIGNER
Eric Williams
DESIGN SCHOOL
Cranbrook Academy of Art
SPONSOR
NYNEX

"This design flawlessly blends a rich palette of imagery, associations and technology to achieve an eye-catching, yet appropriate presence along the roadway...."

JUROR CHARLES BURNETTE, Ph.D.

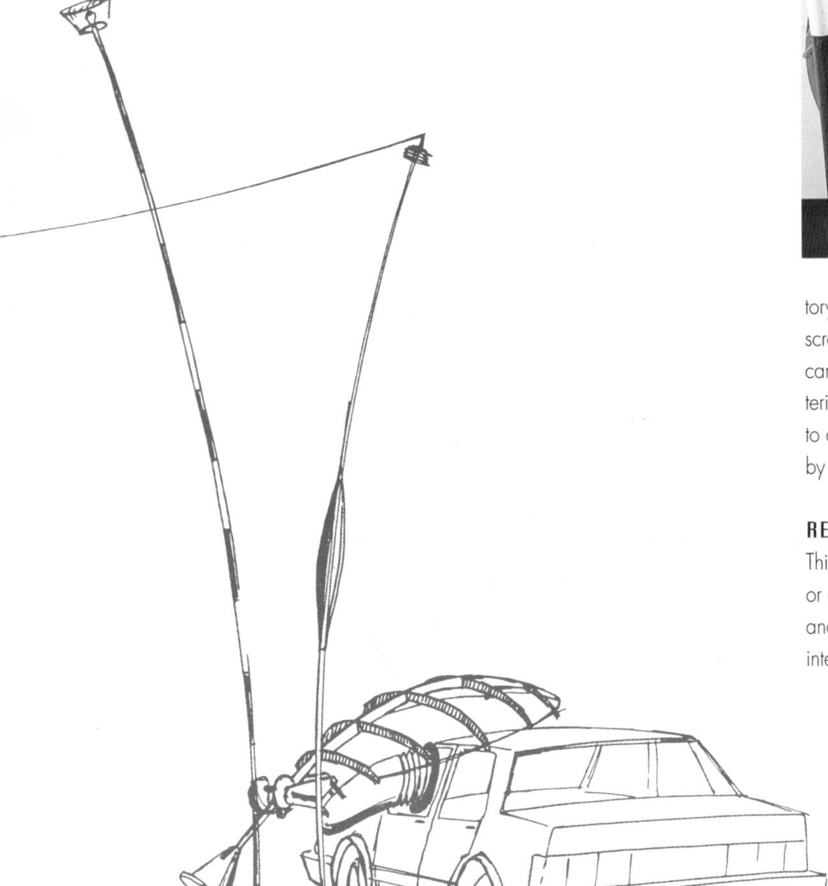

DESIGN OBJECTIVE

Determining which services should be included, whether there should be integration or disintegration of services, public access, privacy, and protocols of use were among the issues that were to be addressed for an advanced generation of public phone booths.

DESIGN SOLUTIONS

This concept for a suburban phone provides drive-up convenience, an electronic directory and a locator function. The information is presented by an LCD screen with a microphone and speaker, which makes contact with the car's window and allows for communication through the window while filtering out extraneous noise. Two vertical elements—a striped signal pole to catch the eye and a transmission pole—are complemented at the base by wheel guides.

RESULT

This exploration showed NYNEX an example of how the situation or context can determine the type and level of service capabilities and equipment, and how its architectural presence can be integrated into the landscape and community.

Retractable Wall Cord

"...How many times at the office and at home have you desperately needed an extension cord? This one is neatly included with your AC receptacles and just waiting to be used..."

JUROR SHARYN THOMPSON, IDSA

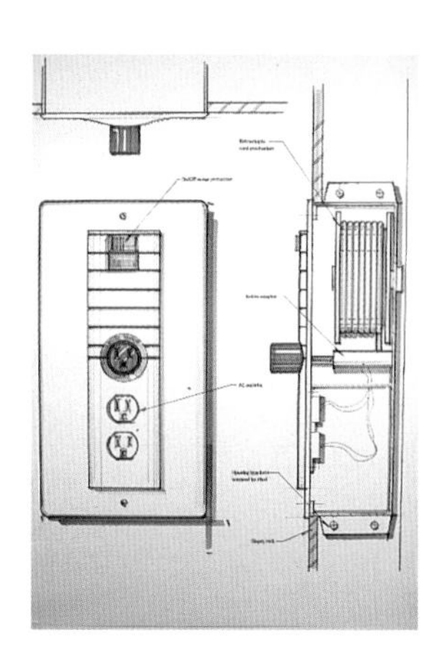

DESIGNERS
Mark Steiner,
Ben Gorbaty of Steiner Design

DESIGN OBJECTIVE
The original concept was to design a retractable cord mechanism that would mount behind the wall.

DESIGN SOLUTIONS
The retractable cord mechanism needed to mount to a typical wall stud of 3½ inches and within the constraints of a ⅜- and ⅝-inch sheet rock dimension. It needed to hold a 12- or 16-foot cord, two AC outlets, a surge protector, reset button and power on/off switch to satisfy UL requirements. In order to fit within these dimensions, the mechanism was positioned parallel to the wall surface, with a roller mechanism facilitating free travel of the cord.

RESULT
The Retractable Wall Cord is an idea, unproved, undeveloped, and unfunded. The product is currently traveling the corporate corridors, searching for funding.

OTHER AWARDS
–Good Design Competition, Chicago

Tutor Training Table System

DESIGN OBJECTIVE

To design a flexible system consisting of one table type and a set of accessory components.

DESIGN SOLUTIONS

Top bridges connect the table tops to form a large number of possible configurations including straight rows, squares, hexagons, or larger concentric rows, while expanding table top area. Basic wire handling in the table itself, augmented by a large wire manager accessory, facilitates the use of computers. A half-round and a trapezoidal table can be grouped with standard straight tables to form a race-track shaped assembly that has an opening down the center to pass wires for electronics. A special mobile unit holds at least eight tables plus the necessary accessories.

RESULT

Early sales figures confirmed that systems sell more units per transaction, and in just a few months the Tutor system pushed sales from zero to 11 percent of total Howe sales.

"...What a wonderful thing it is to see such awareness of detail applied to a product category known for its lack of design finesse."

JUROR VINCE FOOTE, FIDSA

DESIGNERS
Niels Diffrient;
Tom Latone of Design & Development;
Robert J. Ferraro of Howe Furniture Corp.

CLIENT
Howe Furniture Corp.

Concept 2000 Computer System

DESIGN OBJECTIVE

To develop and demonstrate exciting ways Tandem could use its newly developed advanced chip technology in the future.

DESIGN SOLUTIONS

The form expresses advanced structural, cooling and interconnection technologies, and the design provides modular expandability that is much easier and more direct than currently available. Minimal structures reduce costs and complexity and color-coded safety locks and handles enable the user to perform service functions accurately and quickly with little training. Moreover, the components can all be repaired, so they can be reused and reconditioned instead of discarded.

RESULT

Tandem has already begun producing a product with the exposed functional modules that leads the competitors in innovation, configuration versatility, cost efficiency and ease of service.

OTHER AWARDS

–Hannover Fair in Germany, Best of Show

"This design's form literally describes the dual back-up concept that is the essence of Tandem Computers. It breaks the stereotypical computer form by discarding the normal enclosure and expressing the components."

JUROR TUCKER VIEMEISTER, IDSA

DESIGNERS

John Guenther, Brett Lovelady of Tandem Computers Inc.

CLIENT

Tandem Computers Inc.

Biojector

DESIGN OBJECTIVE
To incorporate Bioject's new technology into a rugged, easy-to-clean, compact, and easy-to-operate product.

DESIGN SOLUTIONS
Since the release of air pressure from an internal CO_2 cartridge propels medication through the skin at speeds up to 500 mph and generates considerable heat, the parts had to be able to withstand extraordinary pressure. The cast aluminum body of the instrument functions as a heat sink. The size and placement of the trigger allows individuals with restricted hand mobility to use it.

RESULT
The completed model and production documentation package were delivered within 12 hours of the project deadline. By the year 2000, the annual revenues generated by sales of the Biojector are expected to reach $100 million.

OTHER AWARDS
–*I.D.* 1993 Annual Design Review
–Design Innovations '93 Competition, Design Center, Nordheim Westfalen, Germany, Highest Design Quality Award
–Industrie Forum Design Hannover, 1994 iF Seal for Good Industrial Design

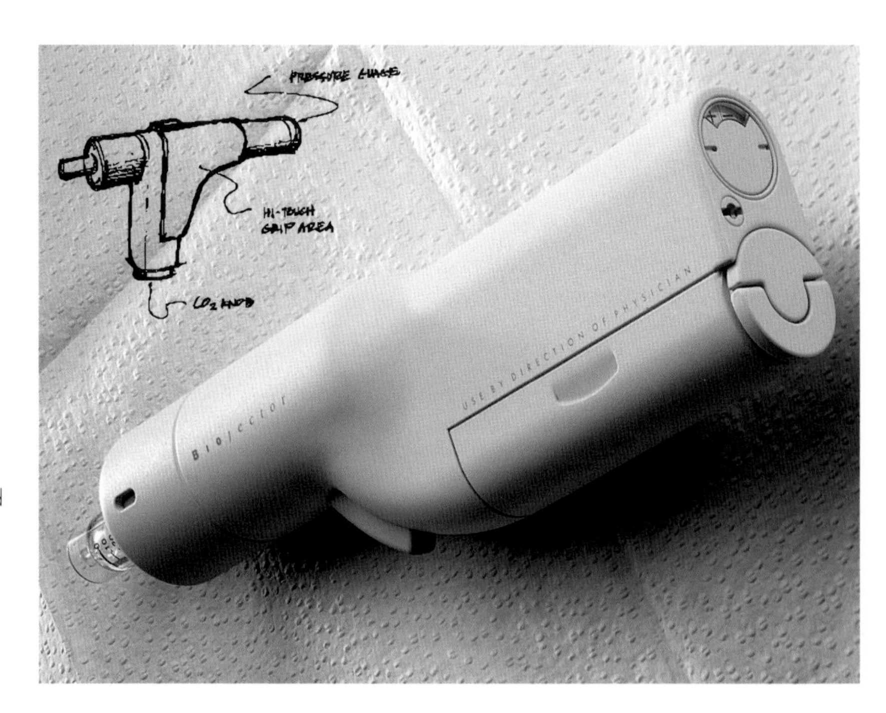

DESIGNERS
Sohrab Vossoughi, Robert Dillon, Ken Dieringer of ZIBA Design; Richard Hollis, Steve Peterson of Bioject, Inc.

CLIENT
Bioject, Inc.

"...[The] plastic medication chamber is the only part of the system that comes in contact with the patient's skin, and it virtually eliminates the risk of cross-contamination of blood-born pathogens to the patient and health care provider."

ADVANCE FOR NURSE PRACTITIONERS, MAY, 1993

Fit

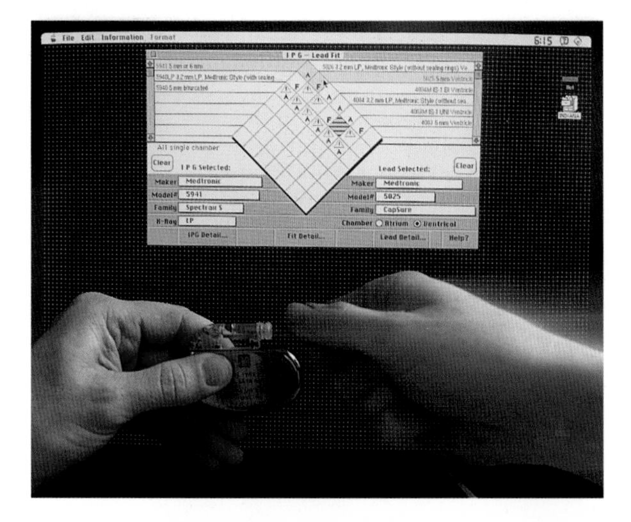

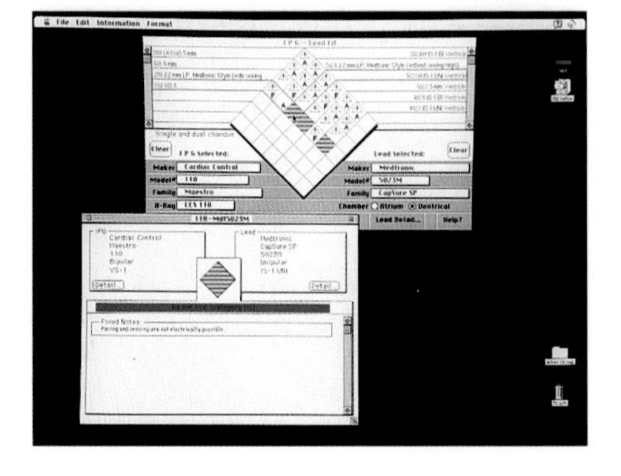

DESIGN OBJECTIVE
A new regulatory standard which would double the already huge database used to determine how to fit pacemakers to pacemaker components required the development of a computer-based system that would simplify access to the information.

DESIGN SOLUTIONS
To best display all of the necessary information, SpectraLogic favored the Macintosh, but since Medtronic had little experience with Macintosh, they wanted to have the actual data located where they could easily maintain it with their IBM-compatibles. So, in addition to the user interface, SpectraLogic built a PC-based file server to store all the data. The software is constructed so that the user can enter any amount of data in any order.

RESULT
The Fit software helps technical support representatives give quick and accurate answers about which pacemaker to use or ways to adapt a pacemaker to a lead when called by surgeons, often during surgery. The value of such accurate and timely information is proven by the 64 percent increase in calls within 12 months of implementation.

DESIGNERS
P.R. Carter, John Oberschlep, David Amis of SpectraLogic, Inc.

CLIENT
Medtronic, Inc.

"...the Fit visual interface design...had to be forgiving as well as fast, with the results very much a matter of life and death. Fit succeeds by providing a beautiful graphic solution to an incredibly complex problem...."

JUROR RITASUE SIEGEL, IDSA

Demon Dispenser

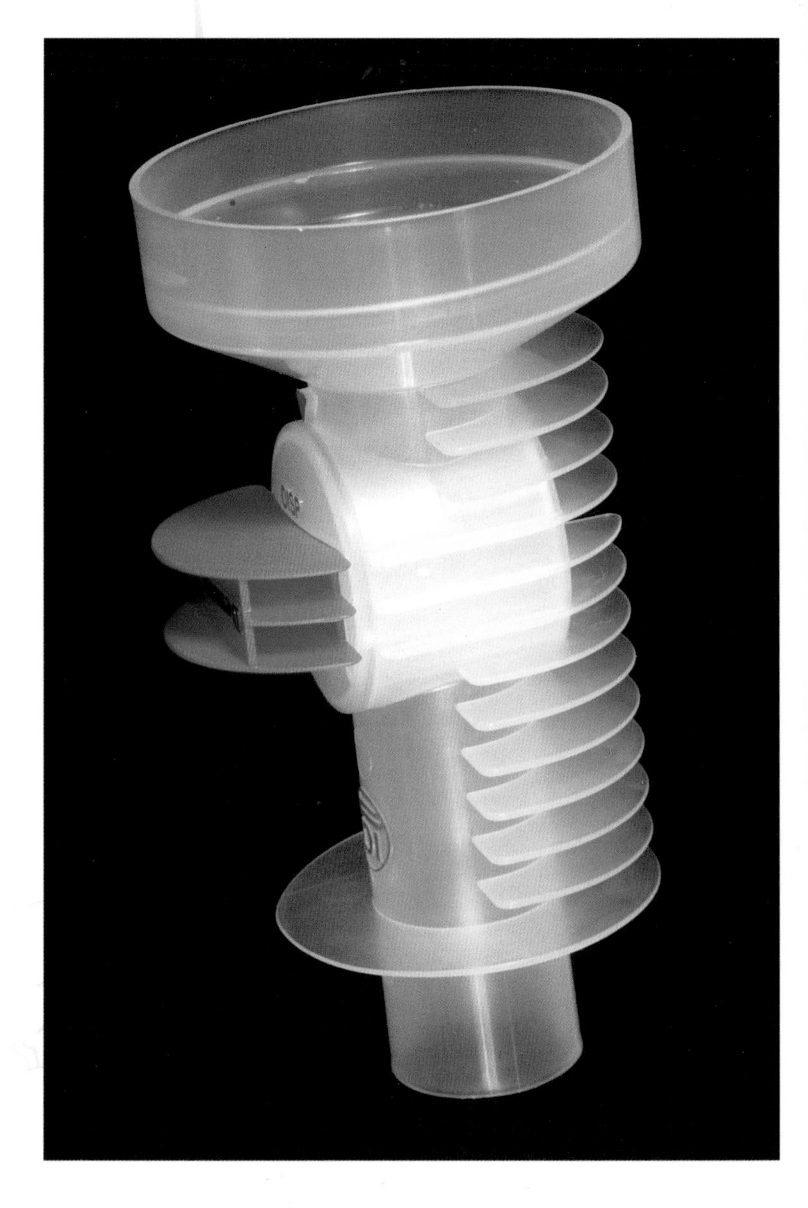

DESIGN OBJECTIVE

To develop a disposable dispenser that would provide the proper measured dosage and reduce the potential negative impact of improper or careless usage of "Demon," a soluble powder insecticide meant to be used in diluted form by an exterminator.

DESIGN SOLUTIONS

Polypropylene was used because of its resilience and extremely low cost, but its warpage and poor tolerancing demanded design ingenuity. The exterior ribs provide critical warpage resistance to the valve cavity. The valve body, insert and handle snap together in assembly, reducing labor costs while enabling removal of the styrene valve. This feature supports material separation and the rinse requirement for federal disposal regulations, facilitates the segregation of materials for recycling, and reduces the risk of misuse as a food dispenser.

RESULT

This dispensing device for concentrated insecticide reduces the user's exposure to the insecticide while it is measured and diluted, providing a foolproof process. Given away free with the insecticide, the dispenser is so well liked that people have been reported stealing them to use with the competitors' concentrates.

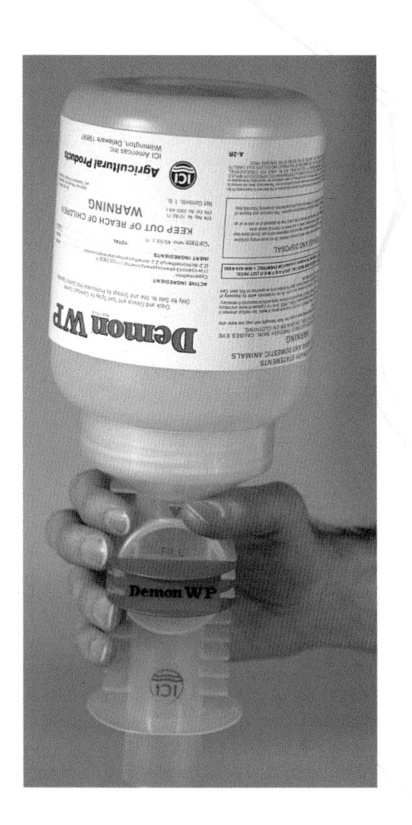

"The game we all play, when asked to design a disposable, is to try to get as much function out of the fewest parts that can be easily assembled at practically no cost. The Demon Dispenser meets all those demands with simplicity. This is truly an elegant solution."

JUROR STEVE HAUSER, IDSA

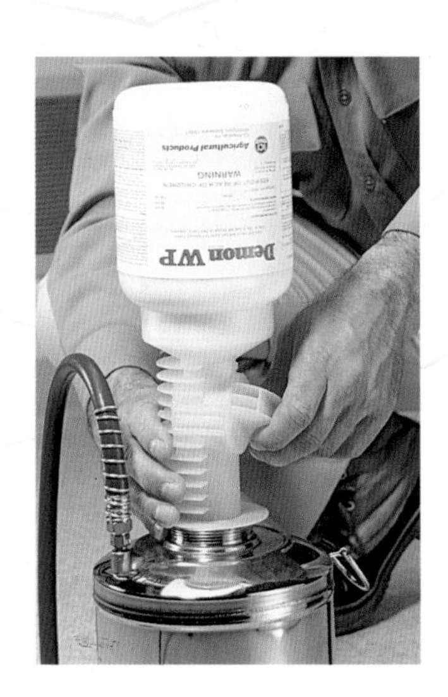

DESIGNERS

Peter W. Bressler, Peter Byar, David Schiff of Bresslergroup; Gerry Gardiner, Tom Stephen, Dick Gouger of ICI Americas, Inc.

CLIENT

ICI Americas, Inc.

"D" Series Angle Wrenches

DESIGN OBJECTIVE

The following requirements were established for the D-Series Angle Wrench: high torque repeatability; a warm, comfortable feel; quiet operation; low torque reaction; and a low-profile trigger big enough to pull with two or more fingers to reduce finger stress.

DESIGN SOLUTIONS

Since the power tools were intended for hand-held use, the functional features and the shape of the housing had to be merged effectively. The symmetrical, reverse valve design gives both left- and right-handed users equal control of the tool, and three different grip sizes (all housed on the tool) accommodate people of all sizes.

RESULT

Careful ergonomic consideration has resulted in a user-friendly design that reduces fatigue and the risk of accidental injury while improving worker productivity. Chrysler Corp.—the wrenches' first major customer—is using the product at its new Jefferson North Assembly Facility, a showcase of worker-oriented processes and equipment.

DESIGNERS
Bob Bruno, Bill Sterling of Group Four Design

CLIENT
Ingersoll-Rand Power Tool Div.

"This design is an outstanding example of concern for the user. Considering the number of people who will use this tool on the production line, this benefit to users stands as an important social gain."

JUROR FRITZ MAYHEW, IDSA

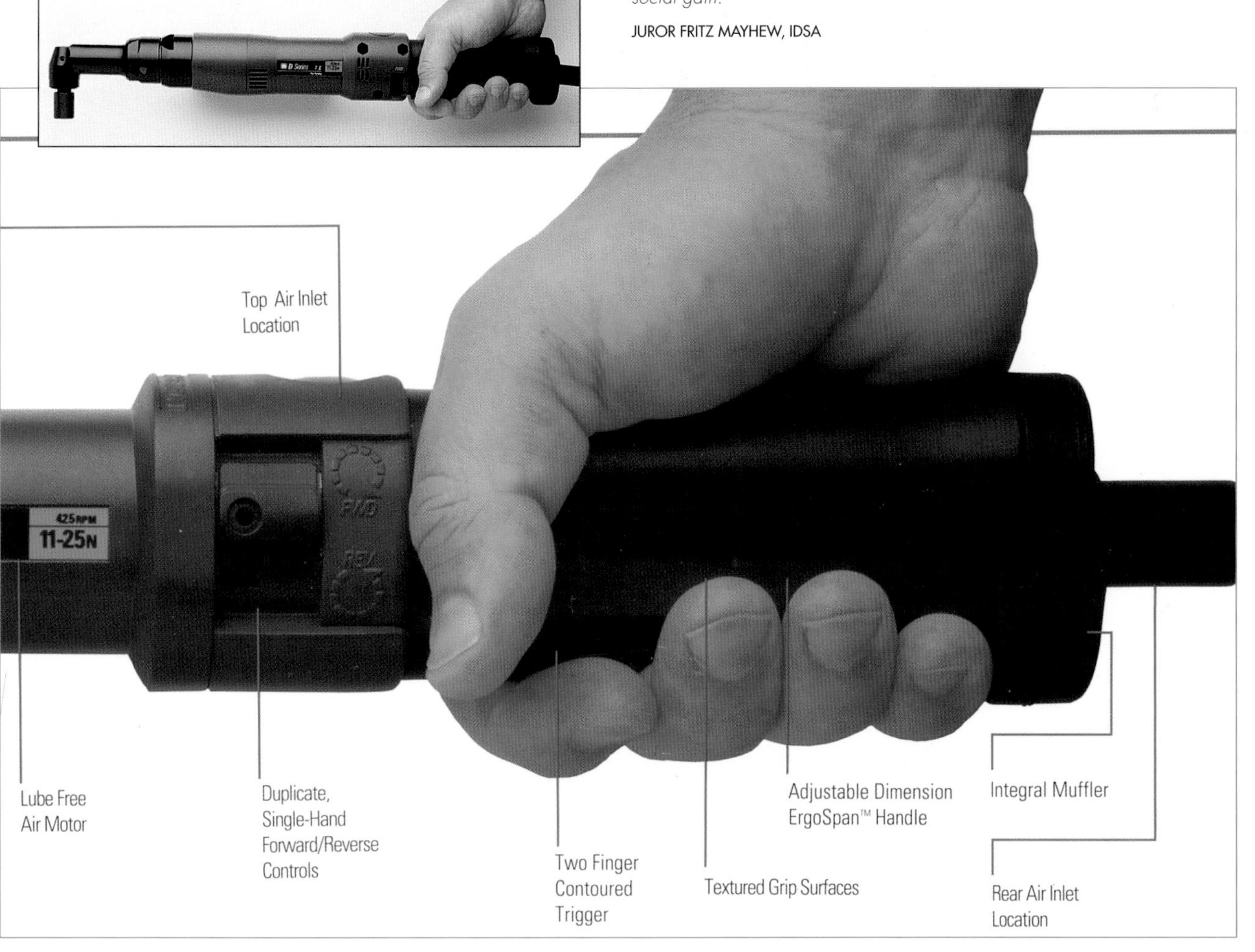

CleanWorks

DESIGNERS
Alex Bally, Gerald Proctor, Charles Kraeuter, Doris Wong of Bally Design Inc.; Allen E. Brandenburg, P.E. of Brandenburg Design and Engineering

CLIENT
Scott/Sani-Fresh International, Inc., a unit of the commercial business Scott Worldwide, Scott Paper Co.

DESIGN OBJECTIVE

To create a simple to operate, safe, environmentally friendly, and cost-effective dispensing system which would replace inconvenient pre-diluted 55-gallon drums with concentrated solution while ensuring accurate dilution strengths.

DESIGN SOLUTIONS

Blow-molded, high-density polyethylene was chosen for the dispenser because this material withstands rough use, is chemically resistant, and requires moderate tooling costs. A Venturi pump activated by the building water pressure aspirates the concentrate and mixes it with the water flowing through the dispenser, providing the correct dilution rate for each chemical. For an instant on/off action, a solenoid valve is activated by mechanically moving a permanent magnet.

RESULT

A compact janitor's workstation, CleanWorks's design keeps janitors out of direct contact with highly corrosive concentrates. The design uses less material than the products it replaces: one six-unit station replaces up to 30 steel 55-gallon drums. Moreover, the plastic components are designed for easy disassembly, re-use and recycling.

OTHER AWARDS

- *I.D.* 1992 Annual Design Review, Gold Award
- *Appliance Manufacturer*, 1992 First Place, Excellence in Design
- Society of Plastics Industries 1992 Annual Meeting, Best of Show Conference Award

"By eliminating the need for 55-gallon drums, the CleanWorks dispensing system reduced the need for on-site storage facilities and improved workplace safety."

I.D. 1992 ANNUAL DESIGN REVIEW

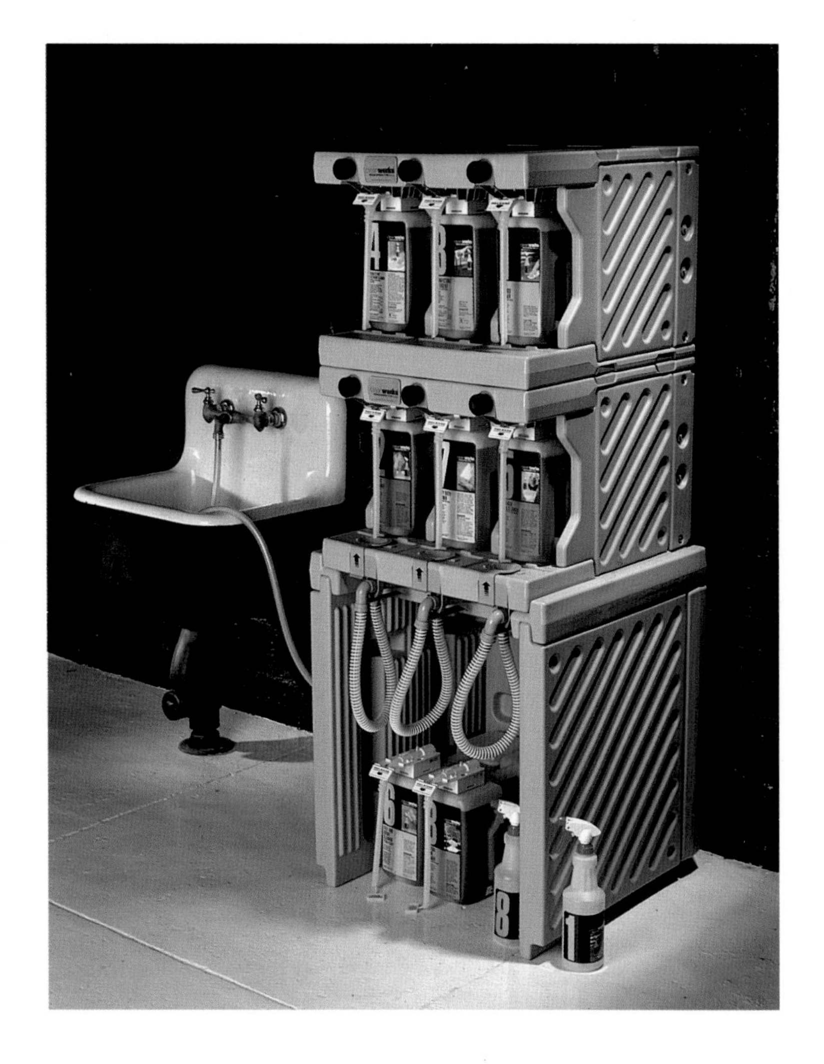

"What we have here is a 'green product' for the commercial market. The metering system ensures accurate usage of materials that are dispensed from reusable, recyclable containers."

GERALD MICHAUD, *APPLIANCE MANUFACTURER*

cleanworks
MODULAR DISPENSING SYSTEM by Scott
RELEASE
DESENGANCHAR
2
GLASS CLEANER
LIMPIAVIDRIOS
8
NEUTRAL FLOOR CLEANER
LIMPIADOR NEUTRO PARA PISOS
5
cleanworks
NON-ACID BOWL CLEANER
LIMPIADOR NO ACIDO PARA INODOROS
READY TO USE
WARNING: KEEP OUT OF REACH OF CHILDREN

Work Group DEChub and DECserver 90L & DECbridge 90

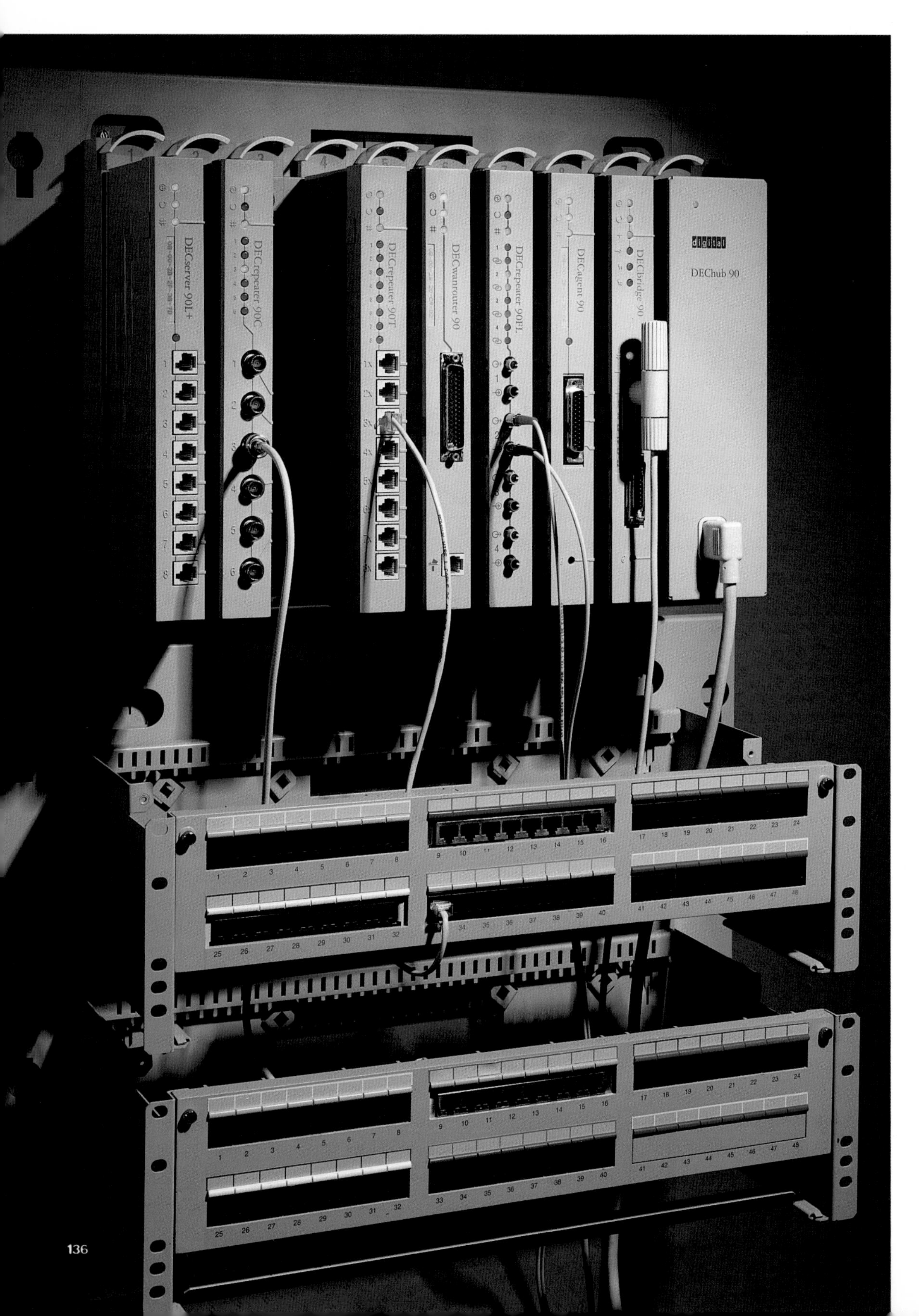

Design Objective

To create a small, low-cost and easy-to-install terminal server for the individual office environment so easy to manufacture that labor costs wouldn't dictate where the product had to be built.

Design Solutions

The system would include a stand-alone terminal server, a work group bridge and a network repeater, all of which would share a common enclosure that could also be used in a hub. A pivot point on the bottom of the hub and a locking finger at the top allows the enclosure to be mounted in the hub with one hand and no tools. To accommodate stand-alone configurations, a rear cover for the enclosure protects the idle hub connector, and includes key holes which allow the user to hang the enclosure on the wall either vertically or horizontally.

Result

The hub is a 10-slot printed circuit board back plane onto which server modules can be snapped. The snap-in design reduces installation time from 1 ½ hours to 20 minutes. An embossed linear pattern represents the internal connectors and LEDs.

Other Awards

–*I. D.* 1991 Annual Design Review

"Users have got to love the plug and play of the new DEC products; the fact that users can set up the system in about 30 seconds is great and the products should sell well on that basis alone."

DOUG GOLD, DIRECTOR OF COMMUNICATIONS RESEARCH AT INTERNATIONAL DATA CORPORATION

"...DEChub is a handsome, tailored, self-assured and authoritative design. It gives us hope that mature functional design, innovative configuration and convivial use of materials might yet rescue the office workplace from the invasion of uniformly anonymous plastic beige boxes."

JUROR ARNOLD WASSERMAN, IDSA

DESIGNERS
Stuart K. Morgan, Meg Hetfield of Digital Equipment Corp.

CLIENT
Digital Equipment Corp.

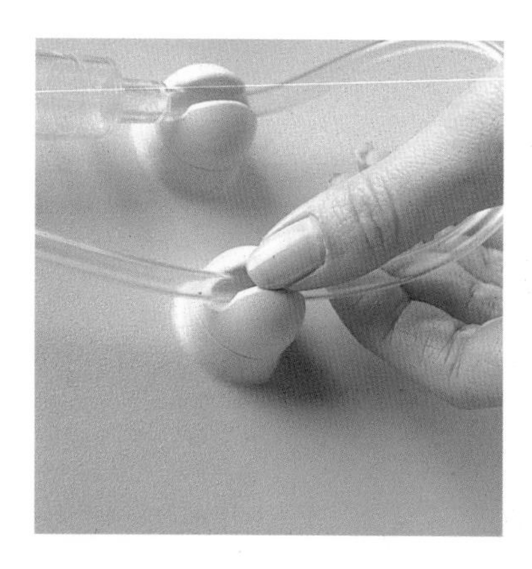

Drake Willock™ System 1000 Dialysis Machine

DESIGNERS
Sohrab Vossoughi, Tom Froning,
Paul Furner, Terry Jones of ZIBA Design

CLIENT
Althin CD Medical, Inc.

"...The jury was impressed by how, through sensitivity to detail and careful organization, this design reduces the threat and elevates the confidence of the patient."

JURY CHAIR JERRY HIRSHBERG

DESIGN OBJECTIVE

To create a dialysis machine with easy set up/cleaning and access to the circuit board that would also reduce a complex and intimidating process into something simple and easy to understand.

DESIGN SOLUTIONS

A confusing array of knobs and controls were replaced with a touch screen. Self-prompting software commands lead the technician through a broad range of treatments eliminating the need for exhaustive training. The confusing crisscross pattern of the blood lines was replaced with a simple linear layout that improves the technician's ability to detect and respond to bubbles in the blood flow. The unit offers fast and easy field service access because each component can be removed without disturbing the others.

RESULT

It takes less than two minutes to access componentry—down from 30 minutes! The new blood line layout reduces the amount of contaminated medical waste by 15 percent, and with unit costs down by 20 percent, significantly higher profit margins have been achieved as well.

OTHER AWARDS

–Design Center Stuttgart,
 1993 International Design Prize
–*I.D.* 1993 Annual Design Review

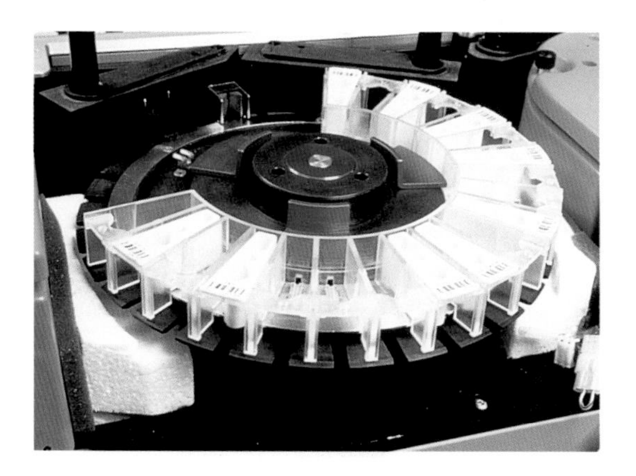

Blood Banking System Research

DESIGNERS
Jane Fulton, Roy Nakashima, Bill Verplank of ID TWO

CLIENT
Syva Co.

"...By analyzing technologists' work habits and building a full-scale mock-up of a blood bank lab, the research method here not only delineated how to structure the hardware, but gave the designers ways to free the technologists from tedious, repetitive tasks, so they can have more time to use their skill and judgment...."

JUROR TUCKER VIEMEISTER, IDSA

DESIGN OBJECTIVE

The Syva Blood Banking System was conceived as an automated blood typing and testing device that would be used principally in hospitals and transfusion centers to determine the ABO/Rh type and screen for antibodies in blood samples. Extensive research regarding current hospital environments, procedures, and staff was required in order to ensure that the Syva system could be successfully integrated.

DESIGN SOLUTIONS

The research consisted of interviews with Syva employees and observations during site visits to blood bank labs and donor centers. Four different methods of research—flow charts, verbal task analysis, storybooks, and full-scale mock-up—offered uniquely useful views of the information.

RESULT

The research identified the points at which errors in patient identification can occur and suggested ways to minimize these errors. Within the blood banks at the hospital, specimens and results were referred to by patient name as well as accession number, but within the donor centers the accession numbers were widely used. Therefore, the system should provide operators the option of linking accession numbers with patients' names. In order to keep the workers actively engaged in their jobs and preserve their sense of responsibility to the patients, the user interface should allow technologists to have ultimate control over how the work is done.

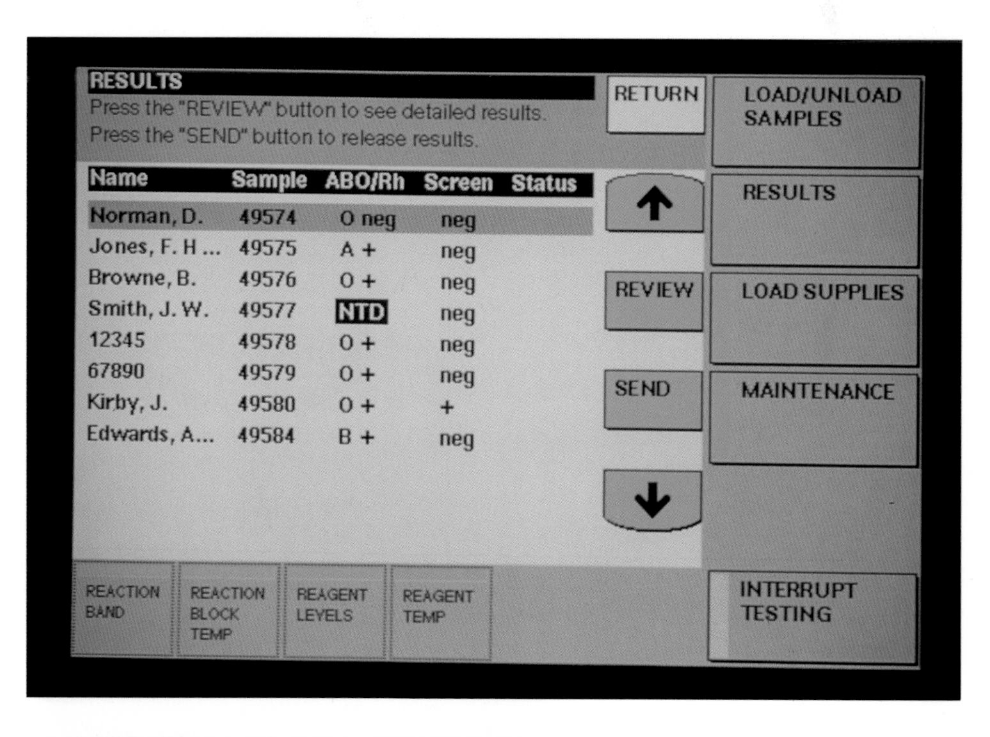

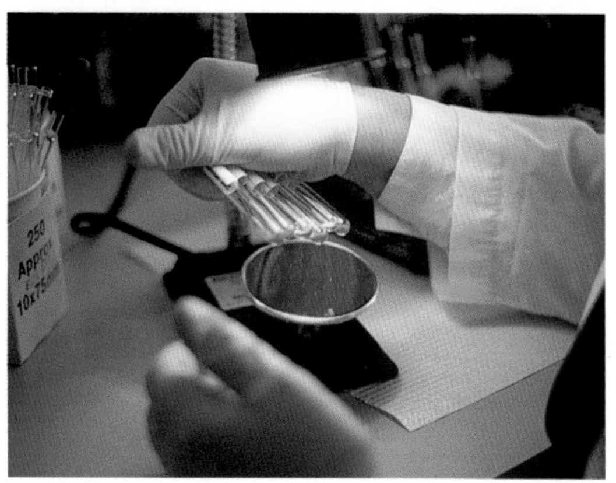

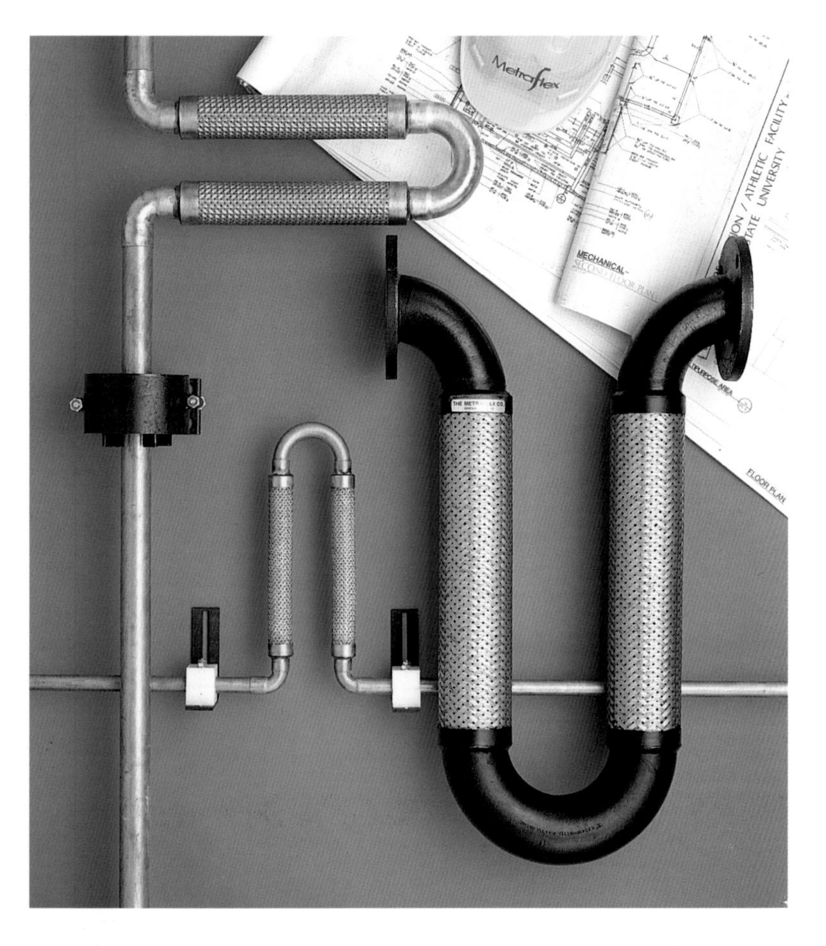

METRALOOP

DESIGNER
James R. Richter of The Metraflex Co.

CLIENT
The Metraflex Co.

The Metraloop allows pipe to flex without breaking during an earthquake. Previous products required three to four times more room than the Metraloop and exerted tremendous loads on the building's superstructure, requiring additional reinforcement. Moreover, the Metraloop requires less labor to install and fewer materials to manufacture than alternative products.

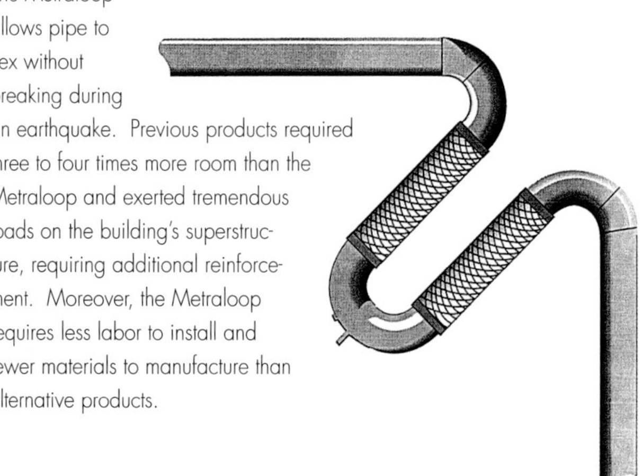

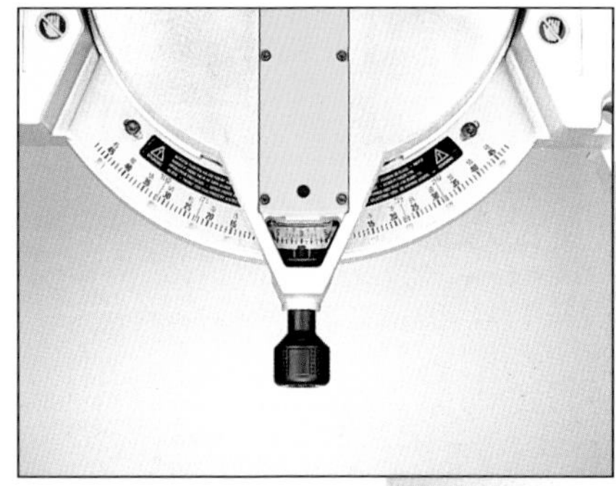

12″ COMPOUND MITER SAW

DESIGNER
Scott Price of Black & Decker Inc.

CLIENT
DeWalt Industrial Tool Co.

Used in the building trade for accurate miter and bevel cuts in wood, plastic and aluminum, the horizontal D-handle has a large three-finger trigger for easy, comfortable operation with either hand. Moreover, the design allows natural wrist movement during operation, reducing fatigue. Made of cast aluminum, it is the lightest large capacity saw currently available.

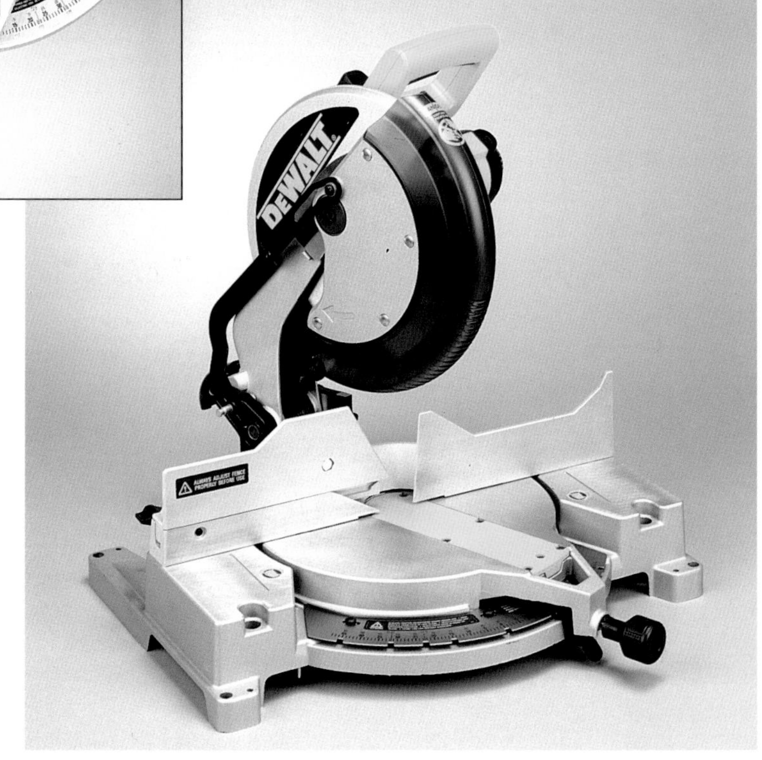

Pressurized Lunar Rover Vehicle

DESIGNERS
Jack D. Lollar, David Hobbs,
Steve Zwonitzer, Jim Catoe

DESIGN SCHOOL
Auburn University

This design would provide astronauts with a safe and comfortable way to navigate the lunar surface. The rover's cross section is elliptical, providing a lower center of gravity, wider wheel base, larger floor space, higher ground clearance and accessible storage space. All exterior moving parts are painted safety orange. The living/lab areas provide multiple crew accommodations.

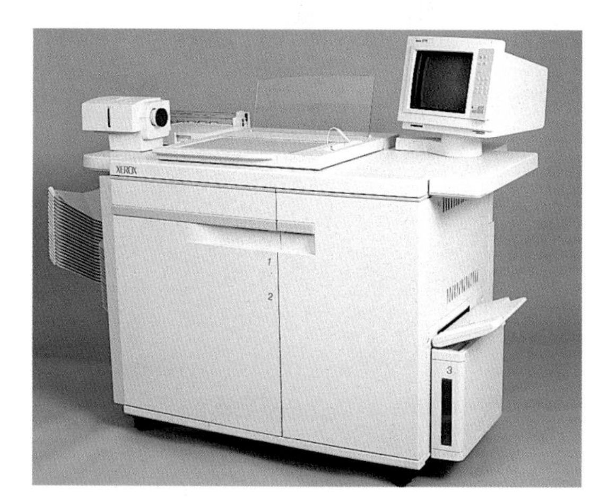

5775 Digital Color Copier

DESIGNERS
Ken Rieck, William R. Hartman of
Xerox Corp.

CLIENT
Xerox Corp.

Although its target market is the traditional printshop operator, the 5775 is a full-color multi-pass copier that features a user interface and activity areas designed to make full-color copying easy for the casual, walk-up operator. The monitor, which displays instructions, can swivel to a position accessible to the wheelchair-based user.

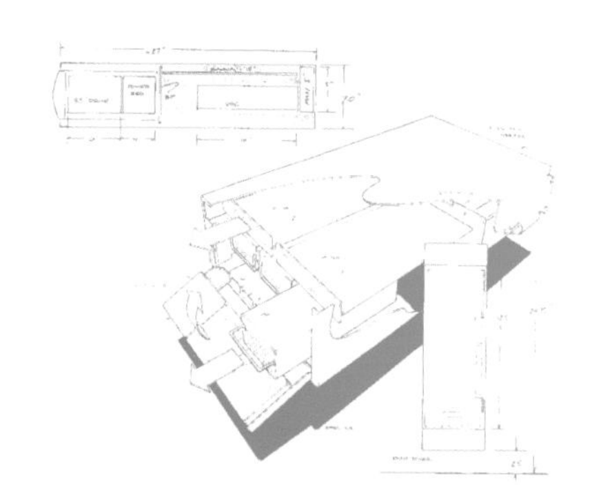

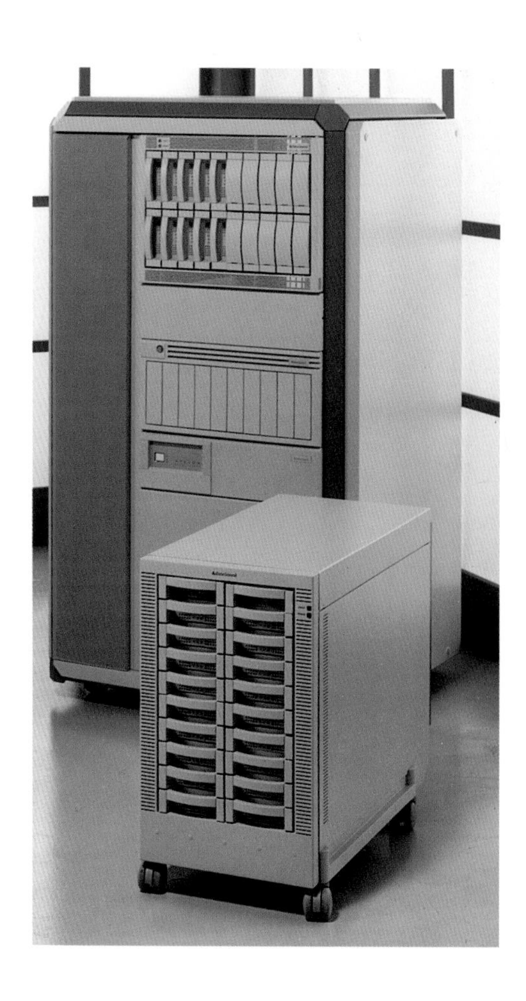

CL2000 Disk Array Office Tower

DESIGNERS
Mark Bates,
Tom Fillio of Data General Corp.

CLIENT
Data General Corp.

The Office Tower succeeds in reducing the equipment's footprint while maintaining easy, tool-free access to all major components. A caster assembly makes it easy to move and doubles as a bumper. The Office Tower parts are made of bromine and chlorine-free plastic, and a 25 percent regrind. No chemicals with CFCs are used in manufacturing.

Series FC Four-Wheel Counterbalanced Sitdown Truck

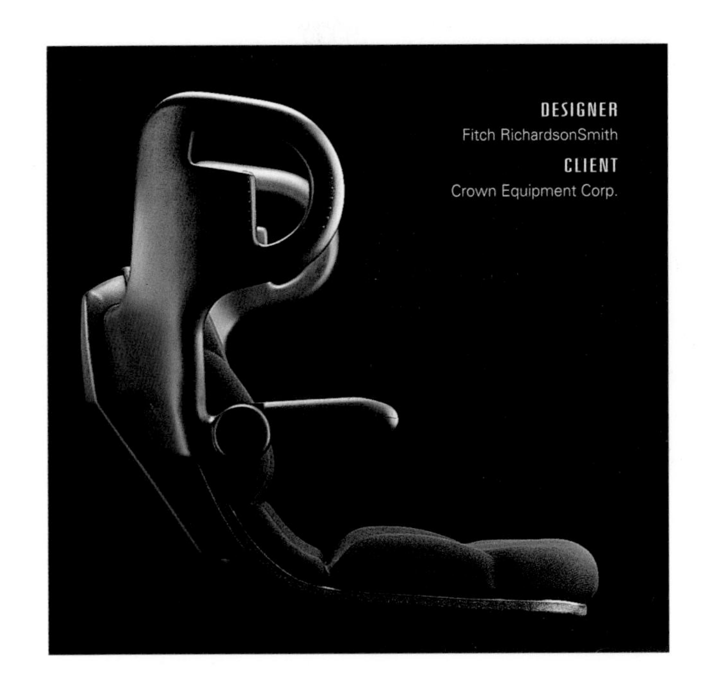

DESIGNER
Fitch RichardsonSmith

CLIENT
Crown Equipment Corp.

Designed to carry between 4,000 and 6,000 pounds of material, this environmentally clean electric lift truck provides good visibility in all directions as well as outstanding human factors in the operator area. The array of controls are laid out based on the natural movements of the human body and offer both tactile and visual feedback. The design makes the truck easy to service and safe to operate.

Ultima 2000 Argon Laser System

DESIGNERS
Max Yoshimoto, Bill Evans,
Gerard Furbershaw, Gil Wong, Keith Willows,
Paul Hamerton-Kelly of Lunar Design;
Peter Keenan, Dave Youngquist, Martha Trupiano,
Greg DuMond, Jeff Saito of Coherent, Inc.

CLIENT
Coherent, Inc.

This argon laser system used in ophthalmic surgery is easy to transport. Among the smallest available, the product's most noteworthy design innovation is the small remote-control pad, which contributes significantly to its flexibility. Its laser and control pad fit together to form a carrying case with telescoping handles and built-in wheels.

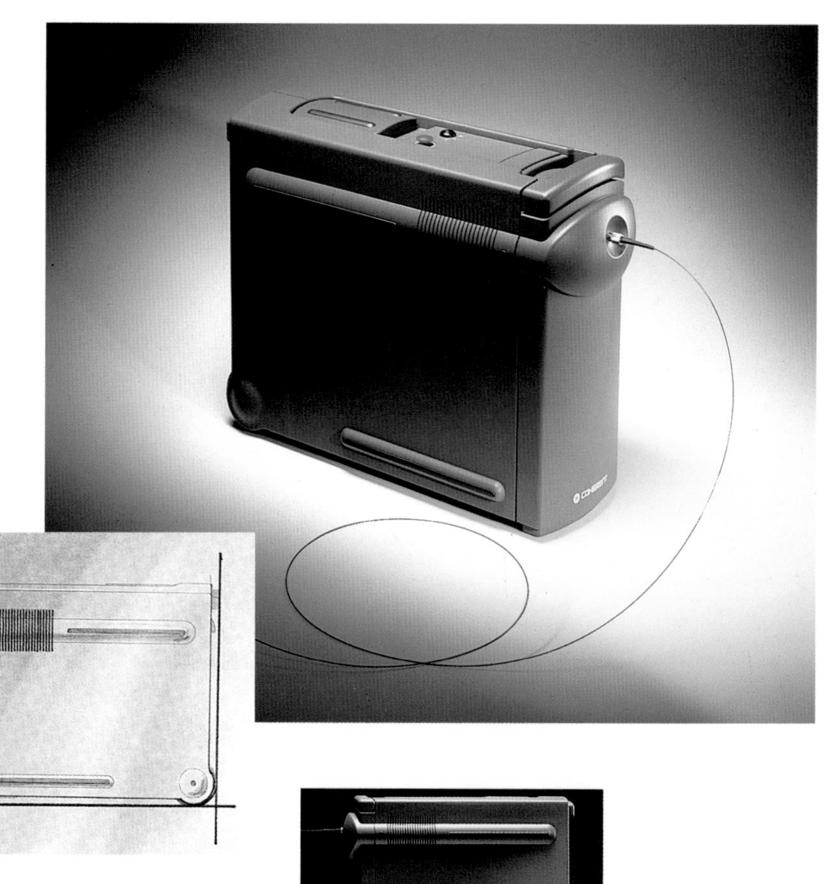

IR2131 Impact Wrench

DESIGNERS
Robert Bruno, David Harris of
Group Four Design;
John Clapp, Robert Geiger,
Pat Livingston of
Ingersoll-Rand Co.

CLIENT
Ingersoll-Rand Co.

Intended for use by professional automotive mechanics, the ½-inch air impact wrench is used to remove and reinstall nuts and bolts. The design's use of space and a composite material significantly reduced the size and weight. The tool's handle—the air flow exhaust—houses an integral swivel inlet, providing quiet operation and insulating the user from cold compressed air.

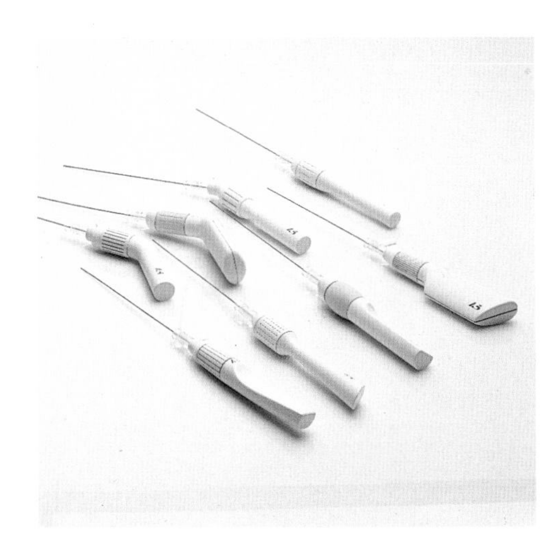

ACS Angioplastic Catheter

DESIGNER
Dan Harden of frogdesign inc.

CLIENT
Advanced Cardiovascular Systems

Angioplastic balloon catheters are used to open blocked coronary arteries. The ACS's innovative shape provides much greater comfort than previous products, minimizing hand and arm fatigue while providing more control and precision. Held between palm and three fingers, leaving the thumb and index finger relaxed and free to turn the wheel, the ACS allows for one-handed operation—a significant improvement.

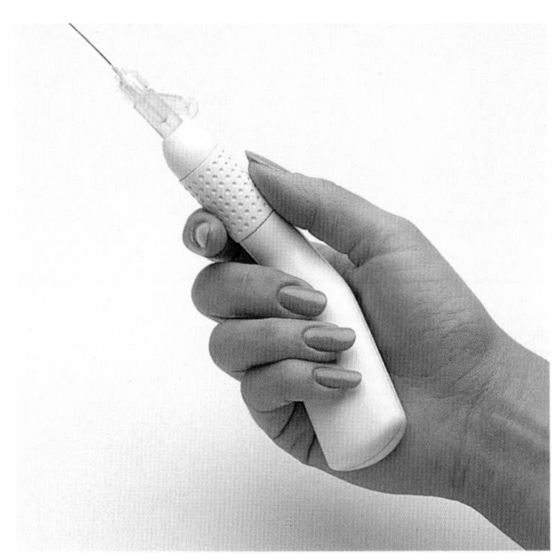

Busse Bac/shield

DESIGNERS
Richard Rosenblatt, MD; John Weaver

CLIENT
Busse Hospital Disposables

The Busse Bac/shield isolates the user in the delivery room from the often deadly viruses and bacteria that can be transmitted while suctioning mucus and amniotic fluid from the newborn infant. The design provides gentle, controlled suctioning, safeguarding the infant from excessive force. The completely sealed one-piece bellows acts as a collapsing piston under vacuum. The walls of the bellows graduate from thinner material at the bottom to thicker at the top to ensure uniform collapsing with suctioning.

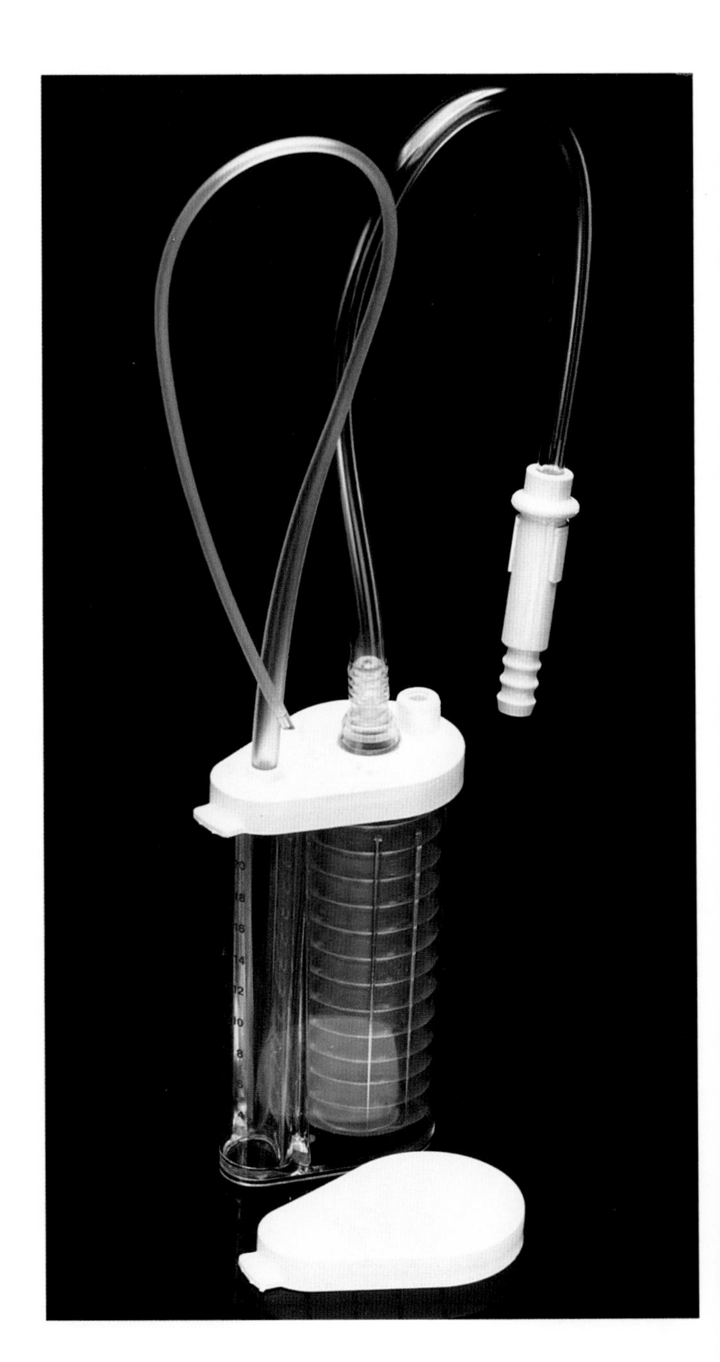

Spectrum Analyzer

DESIGNERS
Sohrab Vossoughi, Mark Stella,
Kuni Masuda of ZIBA Design;
Osamu Komatsu, Masao Komatsu of
Ono Sokki Co., Ltd.

CLIENT
Ono Sokki Co., Ltd.

This hand-held, diagnostic tool measures vibration, sound and frequency in order to identify and quantify defects in large, industrial machinery. Unlike more unwieldy competing benchtop products, this analyzer offers flexibility and mobility. Easy-to-understand icons, a reduced number of operating keys and a large back-lit LCD display make the analyzer simple to use.

IMPRAlert

DESIGNERS
Sohrab Vossoughi,
Tom Froning of ZIBA Design;
Perry Guinn of Impra, Inc.

CLIENT
Impra, Inc.

This surgical and dental safety system detects punctures in surgical gloves to help prevent the transmission of blood-born diseases. A compact device is worn on the user's clothing while skin patches linked by cables detect punctures. A gentle alarm or vibrations alert the user when a puncture is detected. IMPRAlert deliberately resembles a pager in size and design to help users feel comfortable with it.

Future Generation ATM

DESIGNER
Robert Fitzpatrick of
Center for Creative Studies

CLIENT
NCR Corp.

A full-function, self-service ATM machine, this concept features: handicap access; a voice recognition system; exposed details and large wraparound LCD panels. Each input and output slot semantically represents its function—concave forms are used for input slots and convex forms for outputs.

Scale/Printer

DESIGNER
Fitch RS Product Development Group

CLIENT
Toledo Scale

A self-serve scale for use by delicatessen customers, this design integrates a new operator interface into the housing in such a way as to emphasize it as a key feature. A pivot enables the screen to adjust to the height of the user while controlling glare angles. The LCD touch screen eliminates duplicated key functions and allows menu presentation in different languages.

777 Flight Deck

DESIGNERS
The 777 Flight Deck Design Team,
H.G. Stoll, Grace Chan,
Chris Lagerberg of Boeing;
Miguel Remedios of Daedalus

CLIENT
Boeing

The design of this flight deck integrates all the interior components and equipment in an environment that is user-friendly, aesthetically pleasing, spacious and functional. Integration of all the elements reduced the number of parts used while creating an uncluttered atmosphere. Nonreflective, warm tones were chosen for the interior to minimize reflections.

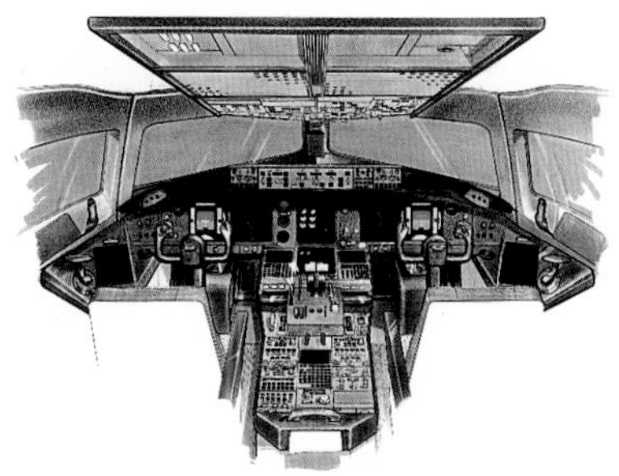

ON.IX

DESIGNER
John Jamieson of Designwerks!

CLIENT
Innovative Technologies

This 386 workstation incorporates an all-in-one, single-board computer within a 12-inch diagonal LCD, dispensing with the need for a CPU entirely. Small in footprint, it replaces the bulky monitor and CPU, taking up little desk space. In fact, the ON.IX even comes with a wall-mount bracket and a touch screen interface. Unlike typical CRTs, the LCD screen gives off no radiation. Moreover, the ON.IX workstation uses low voltage, producing very little heat.

FiberWorld

DESIGNERS
Industrial Design Staff of
Bell Northern Research's Design Interpretive

CLIENT
Northern Telecom

These telecommunications network products simplify operational procedures while meeting stringent technical needs. Concise and organized graphics and the simple, straightforward indicator and alarm strategy reduce human error and confusion. The product often exceeds safety requirements and is designed to ensure that people with less than perfect vision or dexterity will have little trouble operating the products.

High Density Disk Array System

DESIGNERS
Arthur Chin, Ilhan Gundogan of
Data General Corp.

CLIENT
Data General Corp.

This system, with its multiple disks in a single cabinet, provides data storage/retrieval for businesses that cannot tolerate down time. The drives copy data from each other while in use. If a drive fails, an integrated single motion handle allows it to be replaced easily in under a minute. The design also reduced the number of parts and assembly time.

3125 NotePad

DESIGNERS
Werner Stephan of NCR GmbH;
Phoenix Product Design Staff

CLIENT
NCR Corp.

An 80386 computer that recognizes handwritten entries and operates without a keyboard, the 3125 NotePad is the thinnest and lightest notepad available worldwide. Design drove the technology to meet the market requirements of expected size and weight. The battery pack is detachable so that the user can carry spare packs. The 3125 has firmly established NCR's dominance in the market.

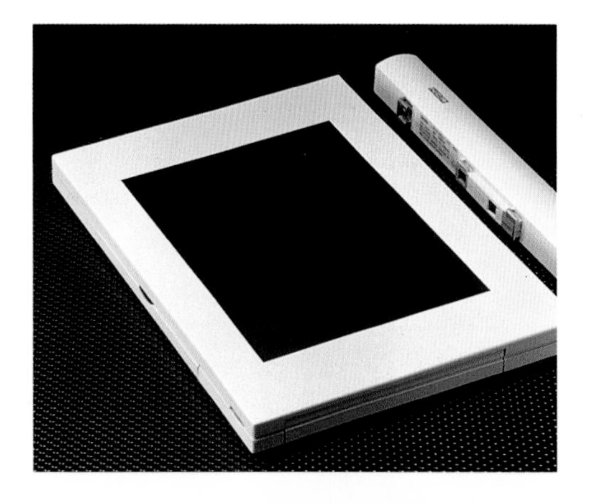

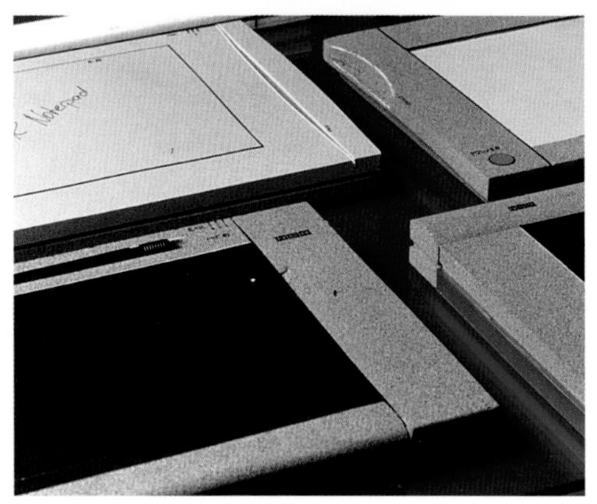

Accumet pH Meter

Used to measure the pH levels, millivolts, temperature and conductivity of liquids, the Accumet meter line includes four models offering a range of sophistication in testing. An estimated 80 percent of the parts are used in all four models.

All the meters are easy to use and clean, and feature an angled clear display.

DESIGNERS
Sohrab Vossoughi, Christopher Alviar, Terry Jones of ZIBA Design; Geoffry Garner, Mitch Houston of Denver Instrument Co.

CLIENT
Denver Instrument Co.

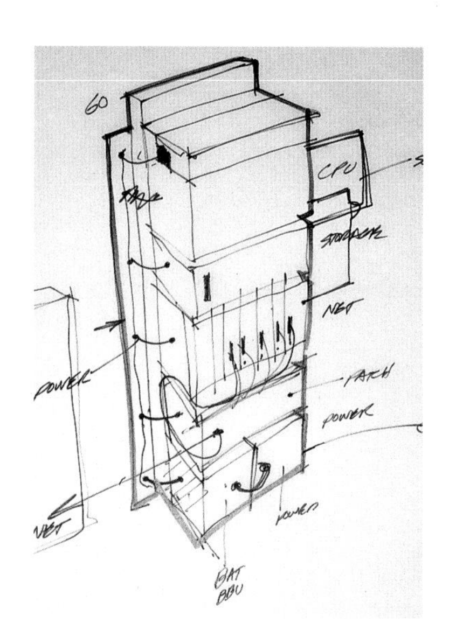

Open Office Enclosure Family

DESIGNERS
Robert Hanson, Meg Hetfield, Scott Baucom, James Walls, David Comberg of Digital Equipment Corp.

CLIENT
Digital Equipment Corp.

These enclosures provide flexible storage for a wide range of rack-mounted computer components. Tall enclosures are approached from a standing position while lower units are meant for seated access. The split door provides easy access and rounded forms make the enclosures approachable.

7780 Workstation

DESIGNERS
Vern Tarbutt, Wayne Fisher of NCR Canada; Graham Marshall of NCR Corp.

CLIENT
NCR Corp.

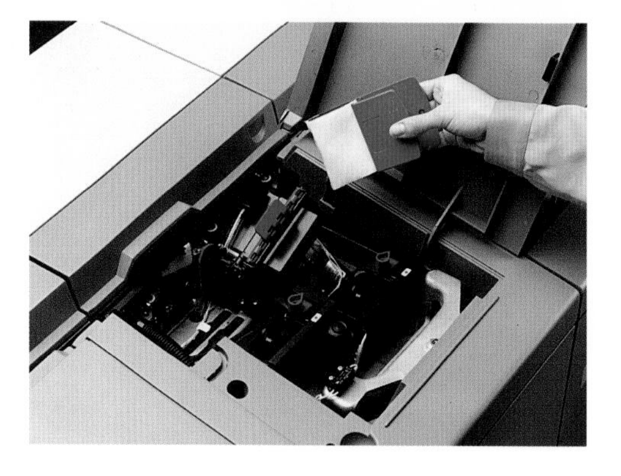

This multifunction workstation serves the needs of three specific check processing market segments: proofing of deposits; remittance processing; and low-speed reading and sorting. In the station's primary operating mode, the user loads documents into the hopper, while interfacing with a computer terminal, and then removes the processed materials from the slots.

Flexible Endoscope

DESIGNER
Robert Bruno of Group Four Design

CLIENT
Schott/Surgitek

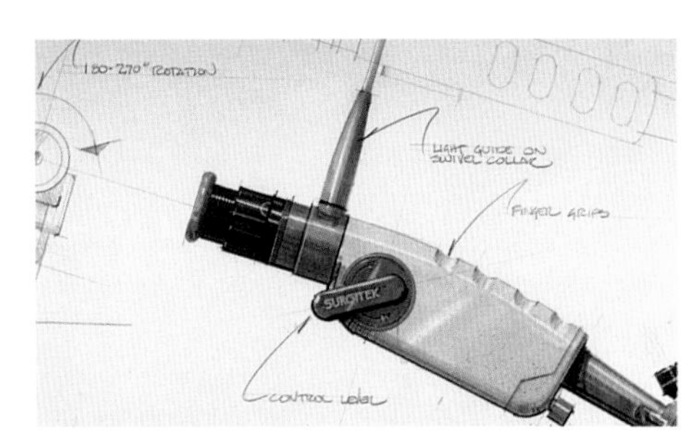

Used to examine and evaluate the state of a patient's urinary tract, the Flexible Endoscope's design reduced assembly time, lowered production costs and is easier to service. The placement and angle of the thumb lever reduces operator fatigue.

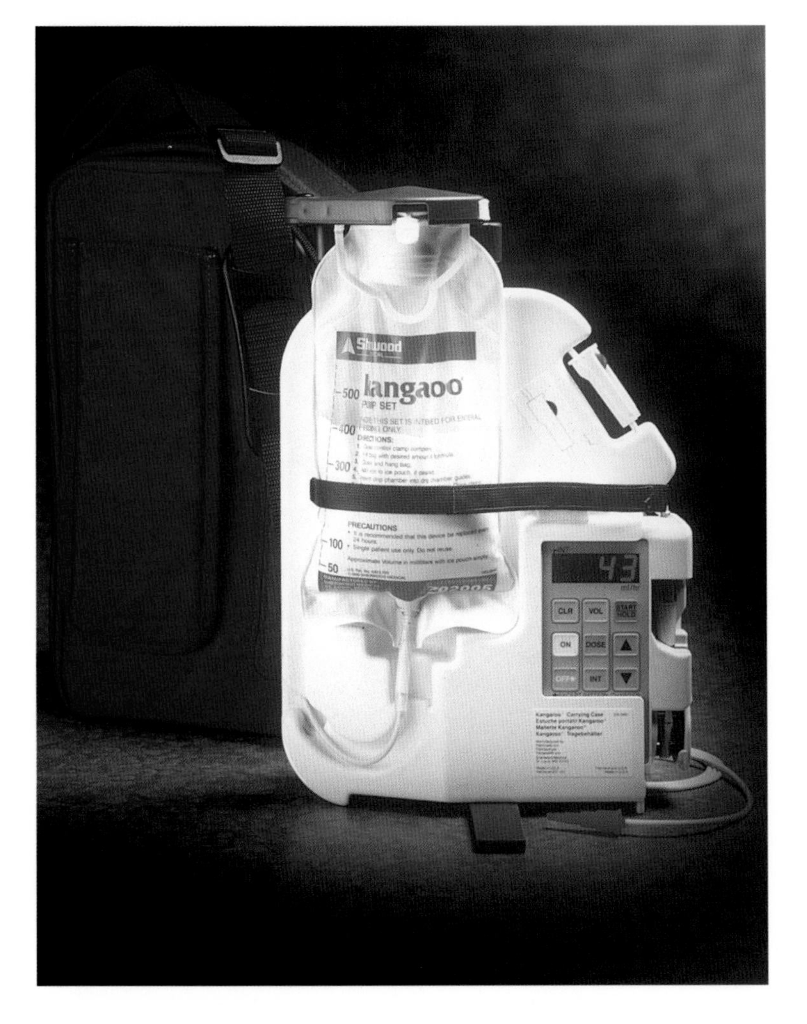

Kangaroo® Pet™ Ambulatory Feeding System

DESIGNERS
Richard A. Sunderland, Clarence L. Walker, John A. Lane, John Allen of Sherwood Medical Co.

CLIENT
Sherwood Medical Co.

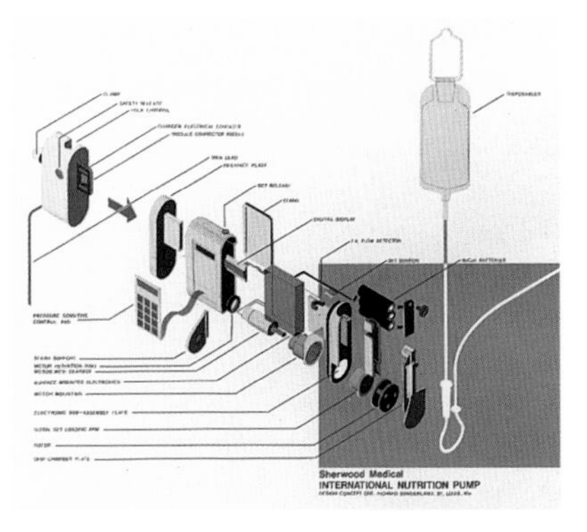

This system allows tube-fed patients freedom of movement while safely and accurately meeting their nutritional needs. The pump delivers liquid food through a tube directly into the patient's stomach at a controlled rate. The unique carrying case is styled to look like an ordinary backpack.

Telesketch

This concept would allow you to draw and design over the phone, sending instantaneous visual information while speaking. The writing palette is transparent, enabling the user to do overlays. Its forms and graphic statements are those associated with traditional telephone design.

DESIGNER
Paul Lanna

DESIGN SCHOOL
Art Center College of Design

Prizm Video Workstation

DESIGNERS
Roy Lonberger, Gary Lancaster,
Paul Rasmussen, William O'Shaughnessy of
Magna Design Corp.

CLIENT
Pinnacle Systems

This workstation's design features a smaller desktop footprint than competitors—a full 486 computer lies in the keyboard—and a significant improvement in ergonomics for the keyboard. The key caps are designed to be the size and shape needed by the predominatly female operators. The design team met the challenge of getting into production within 90 days of the program's start.

Trilliant™ Home Power System

DESIGNERS
Frank R. Wilgus, Lisa Stein, Chuck Leinbach,
Frank A. Wilgus, Dale Benedict,
Inars Jurjans of Fitch RichardsonSmith

CLIENT
Square D Company

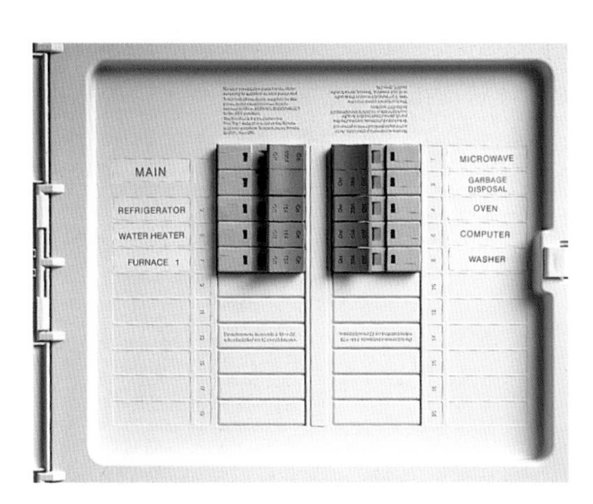

The Trilliant is the first innovation in home load centers in 35 years. Its simple, modular design reduces installation parts from 173 to 42, cutting the installation time by half. One of the best-insulated load centers in the industry, the system has been simplified from a barrage of clamps and wires to four basic elements. Clear graphics communicate a safe, easy-to-operate system.

Oseris 12V/50W Spotlight

DESIGNER
Emilio Ambasz of Emilio Ambasz Design Group

CLIENT
Erco Leuchten GmbH

A 12V/50W low-voltage halogen lamp component, the Oseris can rotate 360 degrees on its vertical axis and uses considerably less energy than equivalent conventional lighting sources. By incorporating a small electronic transformer into the housing, the design reduces the overall number of parts needed for installation, making the system more cost competitive compared to similar systems.

Series 300 IC Wood Sanding Machine

DESIGNERS
Robert J. De brey,
André R. De brey of De brey Design;
Howard Grivna, Gordon Schuster,
Lee Stump, Daniel Hagen of
Timesavers, Inc.

CLIENT
Timesavers, Inc.

Timesavers was loosing its market share to better-designed German and Italian machines. The Series 300 IC Wood Sanding Machine is the designed response, providing much more than an attractive skin to mechanical components. Ergonomic design features include: a perimeter safety rail with a protected, all-around emergency shutdown switch, swing-up belt loading support arms and a front gate safety switch.

International Design and Engineering Assessment Truck

DESIGNERS
L. David Allendorph,
Richard B. Hatch of Navistar International

CLIENT
Navistar International

The IDEA Truck is a functioning concept with real world solutions. Designed to reduce fuel use and address the high driver turnover in trucking, the IDEA Truck features aerodynamic styling that would reduce fuel use by 7 percent. Every surface, part line and detail has a purpose and function. The design provides the driver with comfort, visibility and easy entry/egress.

Adjustable Keyboard

DESIGNERS

Raymond Riley, Dave Shen, Harold Welch of Apple Computer, Inc.; Stephen Peart of Vent Design

CLIENT

Apple Computer, Inc.

This ergonomic keyboard is a success for both the experienced and hunt-and-peck typist. Users can set the angle that is best for them, from completely closed to an open split-wing. The larger spacebar and the palm rests that move with the wings make the keyboard comfortable to use in the open position.

Point of Sale Graphics Research

DESIGNERS

Jane Fulton, Brian Stewart of ID TWO

CLIENT

Spectra Physics

This research investigated the most effective graphic method to communicate proper use for an innovatively shaped bar code scanner. The scanner is vertical, and the best way to scan an item is to move its bar code toward the lower part of the scanner. Through repeated testing with untrained clerks, the team found that it is best to use a graphic that suggests the correct use rather than how the product works.

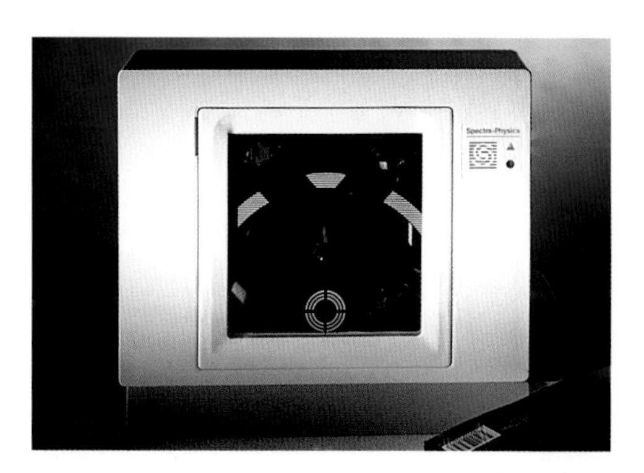

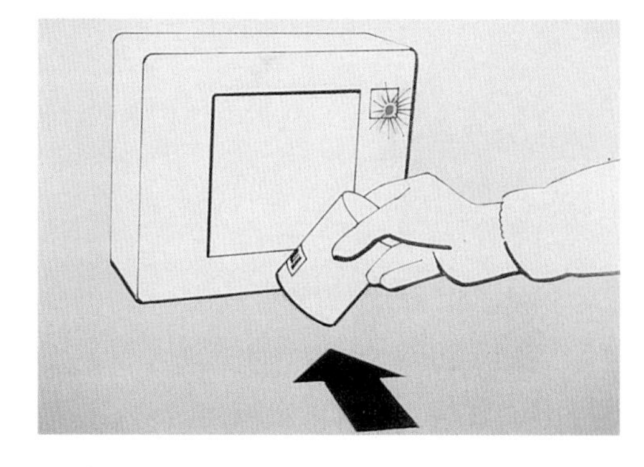

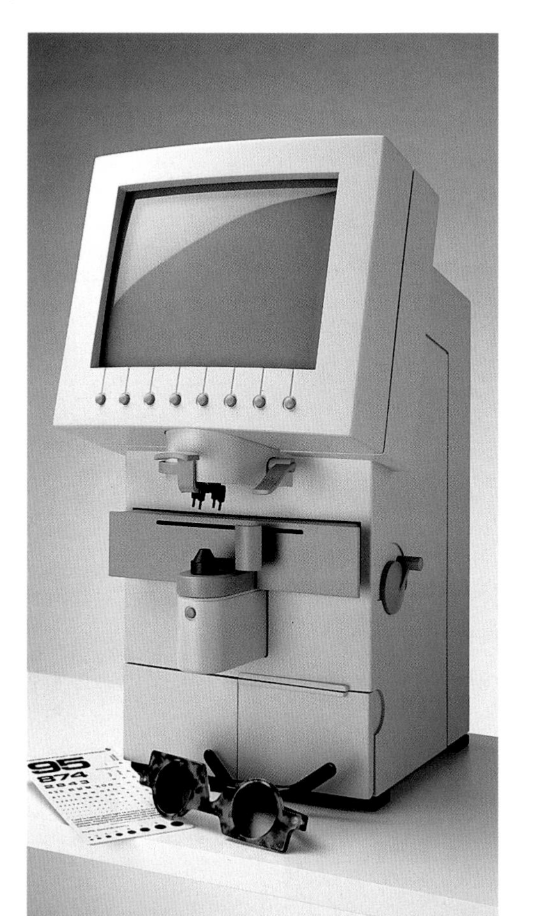

Lens Analyzer 350

DESIGNERS

Frank Friedman, Steve Wittenbrock, Glenn Polinsky of SOMA, Inc.

CLIENT

Humphrey Instruments

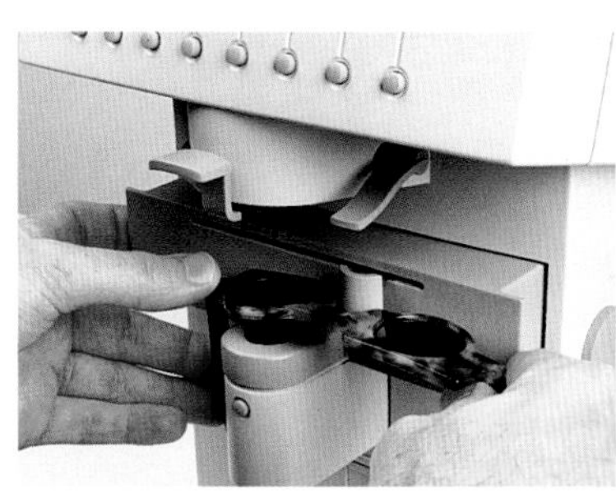

This device guides eye-care technicians through the analysis of glass and plastic lenses. Interaction with the equipment is menu-driven, leading the user through the analysis process. The simplicity and accuracy of the analysis process eliminates the need for extensive training.

In Search of an Unsurpassed Reputation for Design Excellence

DESIGNERS

Bryce G. Rutter, Ph.D. of
Metaphase Design Group Inc.;
Gary N. Wagner of NCR Corp.;
Stephen Wilcox, Ph.D. of Design Science

CLIENT

NCR Corp.

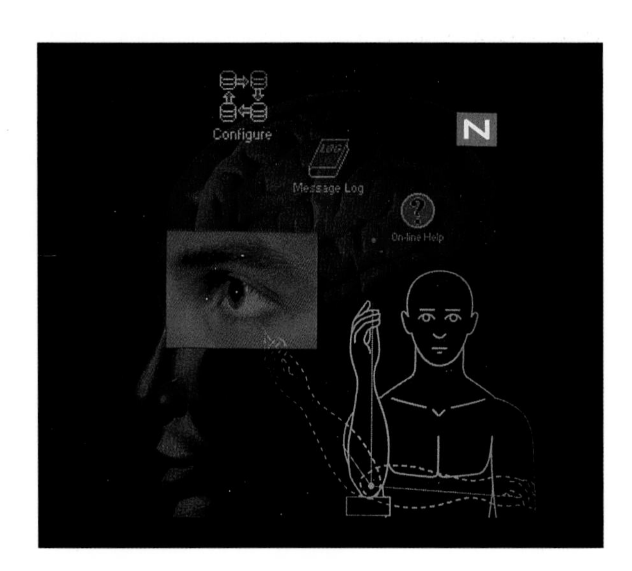

This research defined "What is design excellence?" using database searches, competitive product analysis, personal interviews with US and foreign design leaders and customers, and qualitative and quantitative studies of various manufacturers' products and corporate characteristics. The research found that design excellence does not exist in a vacuum. The results have provided NCR with a foundation on which to construct its current strategic plan.

The Search for Design Excellence

R&D and Profitability

Original Ranking	R&D costs 1990÷1986 (Factors)	Gross sales 1990÷1986 (Factors)	Avg. Net Income 1986-1990(wtd.) (in millions $)	1990 R&D Costs (wtd.) (in millions $)	Net Income 1990÷1986 (Factors)	Avg. Profitability (1986-1990)
A	3.10	2.92	340(3400)	478(4708)	3.08	8.64%
B	1.64	2.41	408(1046)	1153(2956)	3.02	2.78%
C	1.65	1.86	708(2529)	1367(4882)	1.43	7.21%
D	6.98	5.75	244(4067)	185(3083)	10.6	10.96%
E	1.69	1.53	569(837)	1853(2725)	2.69	2.14%
F	1.98	1.71	841(3004)	1614(5764)	.12	8.12%
G	?	1.90	40(5714)	?	.65	11.1%

Original Rankings: rankings as determined by project team
Factors: a number that multiplied by the 1986 figure equals the 1990 figure

Spectrum NDT Transducer

DESIGNERS

LeRoy J. LaCelle, Kurt Solland, Tony Grasso of Designhaus, Inc.

CLIENT

Staveley Sensors, Inc.

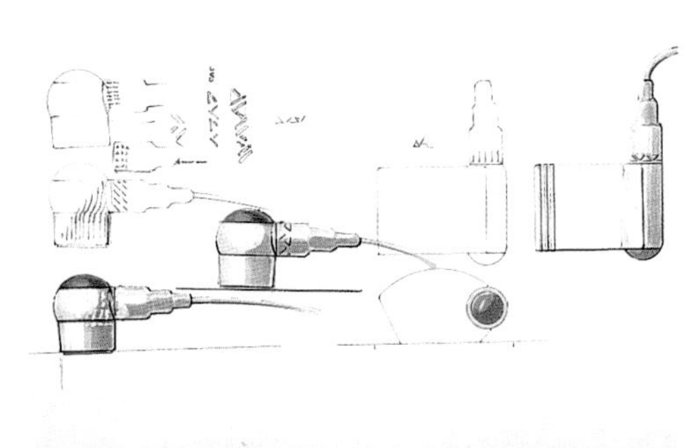

This nondestructive testing transducer provides a comfortable form for users in a rough and often outdoor industrial environment. Extremely durable, the ball is a comfortable interior palm shape that allows the user to accurately position it under various surface pressures. The color cap easily identifies the unit's frequency performance, and an integrated connector jacket seals the electronic interface from the environment.

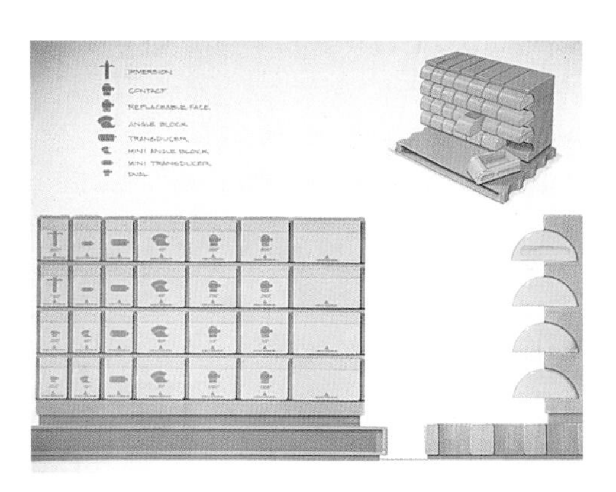

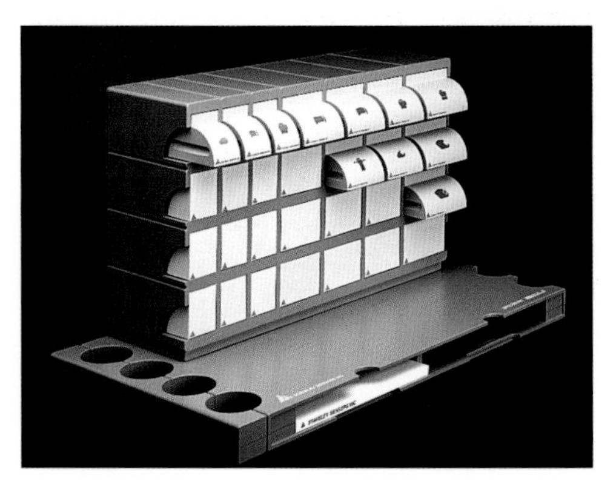

30 Series Clampmeters

DESIGNERS

Indle G. King, George L. McCain, Jeffrey C. Brown of Fluke Corp.

CLIENT

Fluke Corp.

These clamp meters are used to troubleshoot electrical power distribution systems by clamping around insulated wires and displaying information on the current in the wire. The rugged, angular shape of the jaws allows the electrician to pry apart wires in a bundle without touching the wires, and the handguard provides a barrier, as well as a visual clue, to the zones that are safe to handle. It is the first product line in its class to incorporate a physical guard as a safety feature.

DataVault and Connection Machine-2A

DESIGNERS
Marc Harrison, Mark Barthold of Marc Harrison Associates; Sheryl Handler, W. Daniel Hillis of Thinking Machines Corp.

CLIENT
Thinking Machines Corp.

The DataVault provides mass storage for Thinking Machine's parallel processing supercomputers, and the Connection Machine-2A is a reduced-sized version of Thinking Machine's large devices. Their design solves cooling, manufacturing, set-up and service problems simply and makes an effective aesthetic statement.

Top cube elevates
for access to Core.

Unconventional
air-flow cooling

Computer Core

fan-deck

Power Supplies
Inputs/Outputs

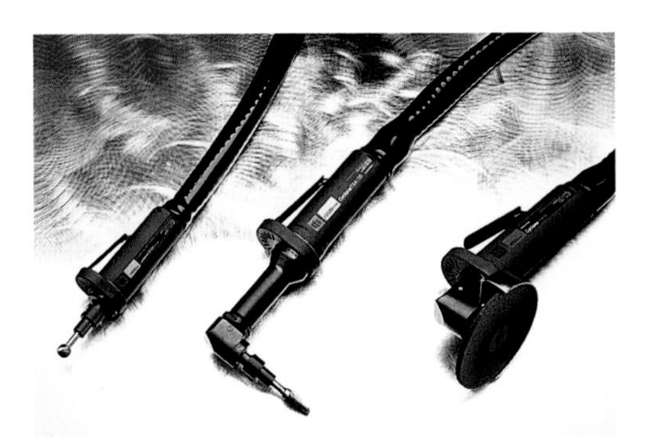

Cyclone Grinders

DESIGNERS
The Group Four Design Team;
The Ingersoll-Rand Cyclone Development Team

CLIENT
Ingersoll-Rand

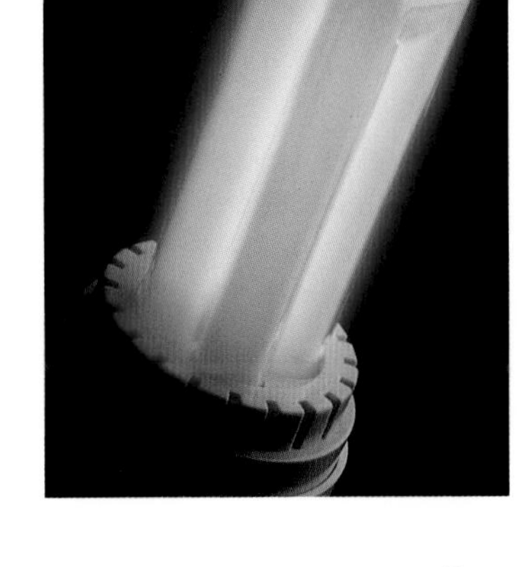

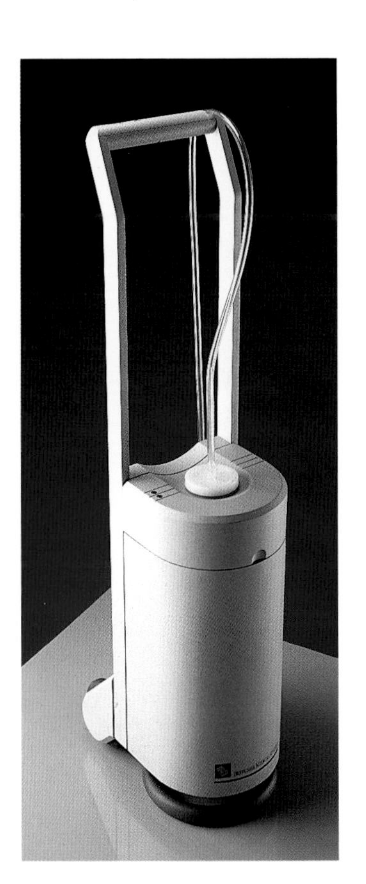

Smoke Evacuator

DESIGNERS
Allan Cameron, Douglas Walker,
Kitch Wilson of S.G. Hauser Associates, Inc.

CLIENT
Birtcher Medical Systems

Fluorescent Lighting Products

DESIGNERS
Peter Lowe, Peter Muller, Randy Bleske,
Peter Edwards, Kathryn Sakata of Interform

CLIENT
Lumatech

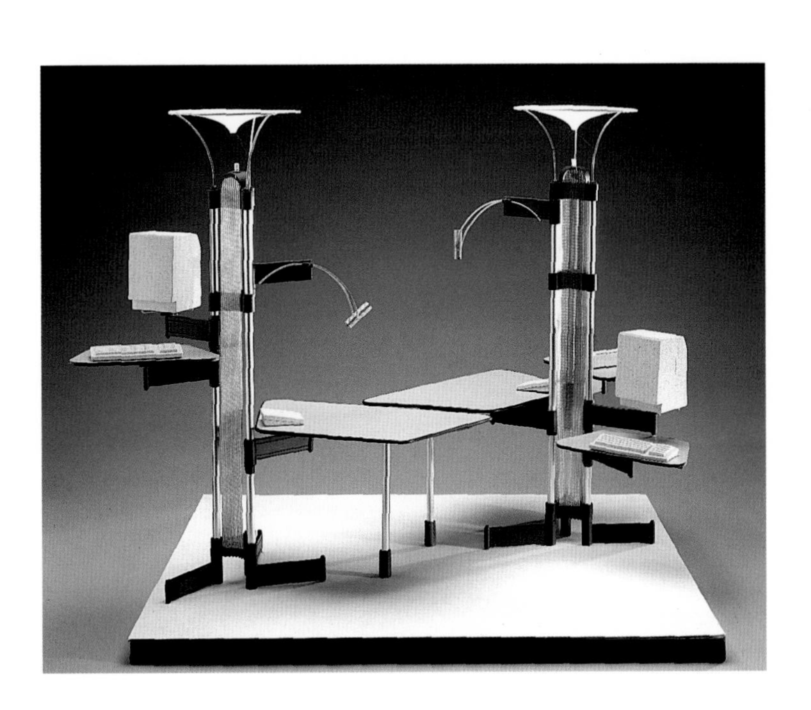

Component Tree

DESIGNER
Peter Arnold

DESIGN SCHOOL
Art Center College of Design

SPONSOR
Steelcase

Insyte® Saf-T-Cath

DESIGNERS

Timothy J. Erskine,
Gerald H. Peterson, E. Robert Purdy,
The Product Development Team of
Becton Dickinson Vascular Access

CLIENT

Becton Dickinson Vascular Access

AccuPinch

DESIGNERS

John von Buelow, Mark Andersen,
Richard Spindel, Martha Stock, Sam Iravantchi of
S.G. Hauser Associates, Inc.

CLIENT

Hycor Biomedical Inc.

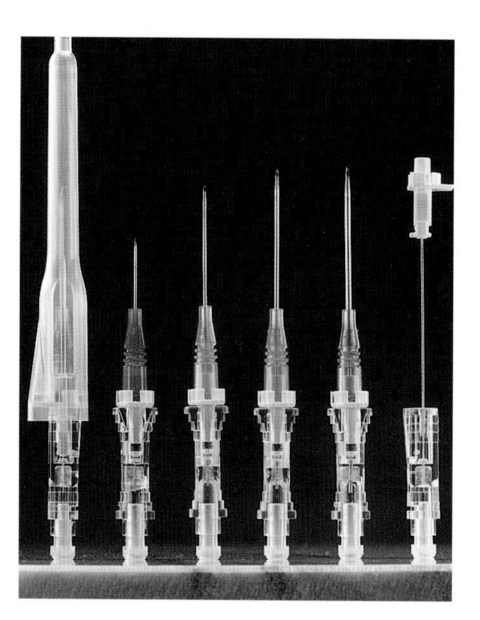

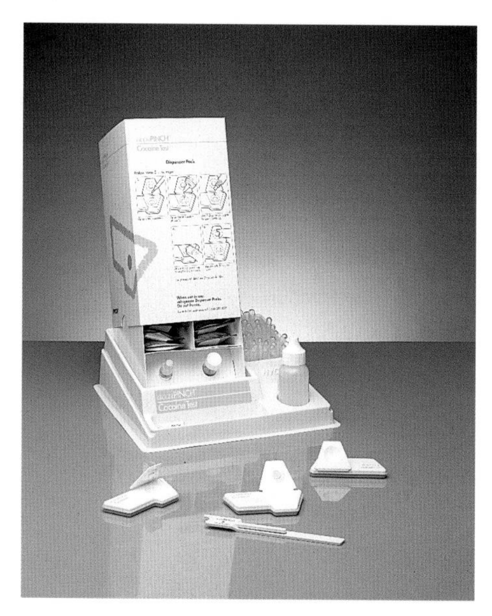

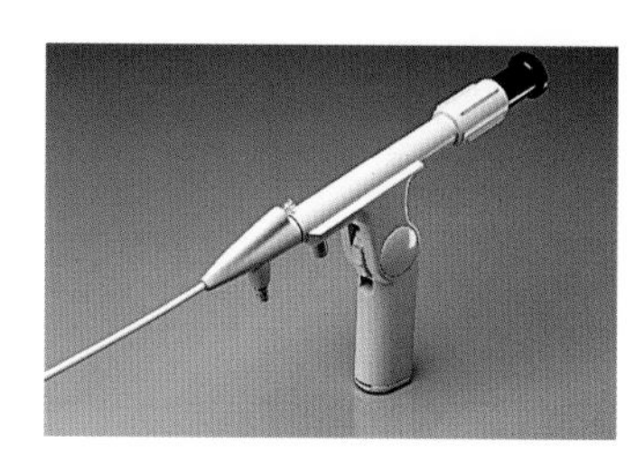

Endoscope

DESIGNERS

Steve Montgomery, Paul Gaudio,
Mark Andersen, Karel Slovacek,
Dave Stocks of S.G. Hauser Associates, Inc.

Millipore Third World Environmental Incubator

DESIGNERS

Matthew Bantly, Kathryn McEntee
of Roche Harkins Inc.

CLIENT

Millipore Corporation

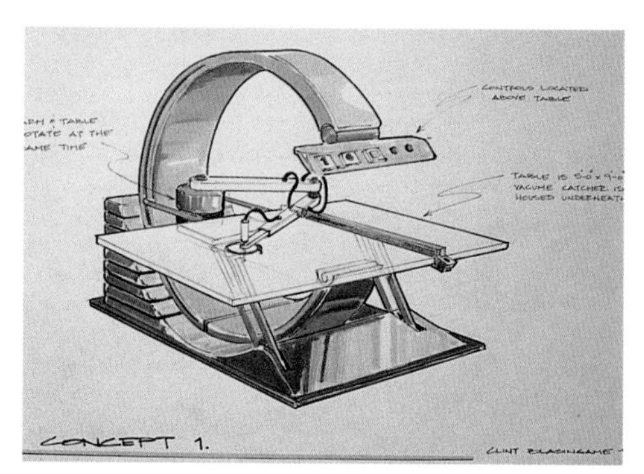

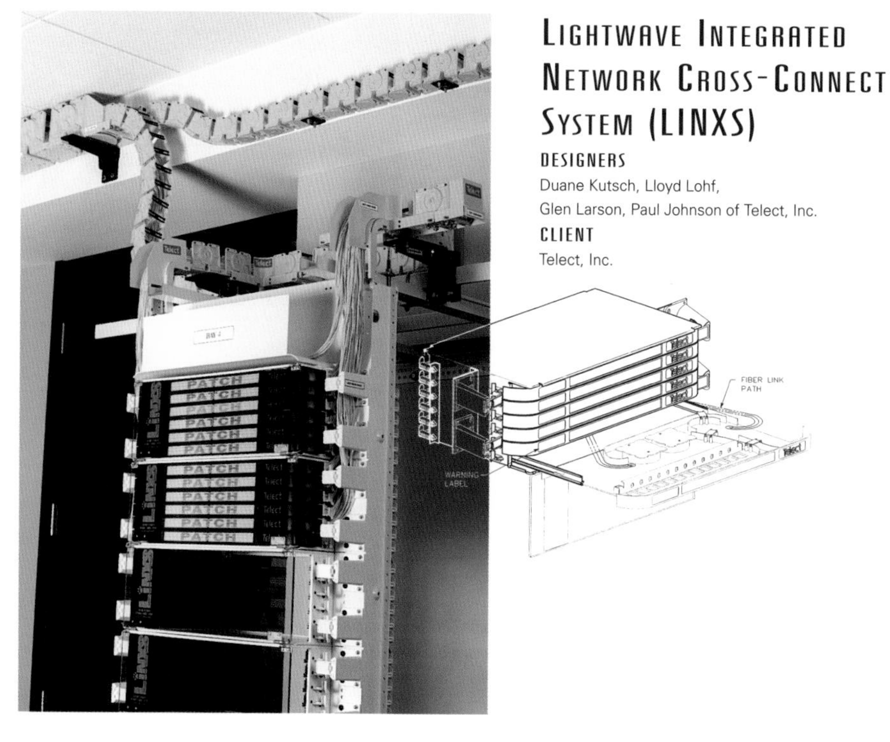

Lightwave Integrated Network Cross-Connect System (LINXS)

DESIGNERS
Duane Kutsch, Lloyd Lohf,
Glen Larson, Paul Johnson of Telect, Inc.

CLIENT
Telect, Inc.

VersaJet Water Jet Cutting System

DESIGNER
Clint Blasingame

FACULTY ADVISOR
Irek Karcz

DESIGN SCHOOL
Auburn University

TestBook

DESIGNERS
Gilbert Lemke,
David Skinner of
Hewlett-Packard

CLIENT
Hewlett-Packard

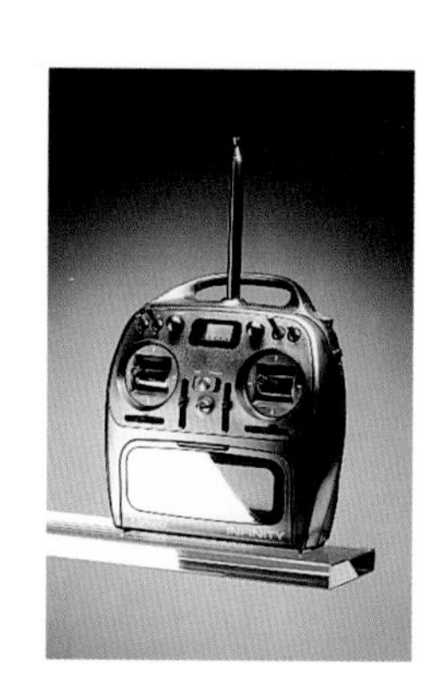

Infinity Radio Control

DESIGNERS
Fernando Pardo, John Cook,
James E. Grove of Designworks/USA;
Barbara Renaud, Tim Renaud,
Bob Renaud of Airtronics, Inc.

CLIENT
Airtronics, Inc.

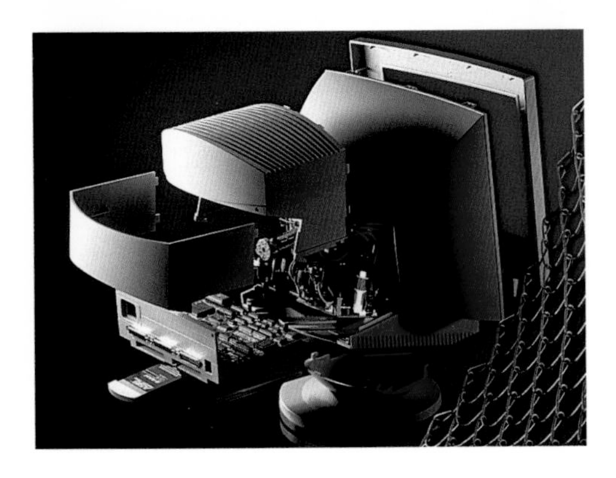

440 Personal Communicator

DESIGNER
the frogdesign team

CLIENT
EO Inc.

MAX/IT™ Desktop Terminal

DESIGNERS
Edward Cruz, David Weir, Vincent Razo of S.G. Hauser Associates, Inc.;
David Brooks, James Jan of Link Technologies

CLIENT
Link Technologies

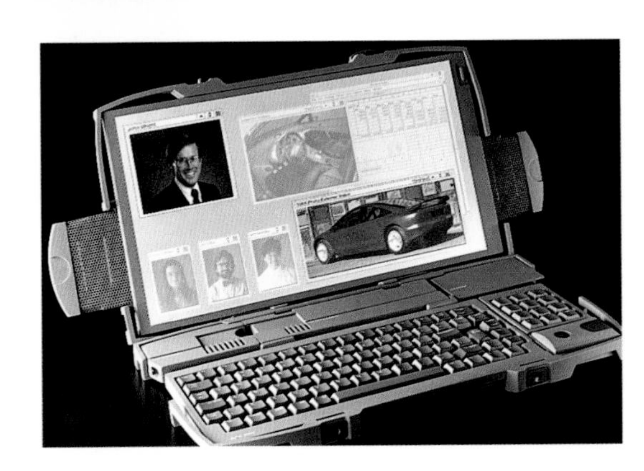

Portable Multimedia PC

DESIGNER
Robert J. Shapton

DESIGN SCHOOL
Center for Creative Studies

SPONSOR
NCR Corp.

PoqetPad

DESIGNERS
Robin Chu of Taylor & Chu;
Mark Schiebl, Chris Laing of Poqet Computer Corp.;
Katsuyuki Fukutake, Masahara Kasahara, Shigeru Hidesawa of Fujitsu

CLIENT
Poqet Computer Corp.

Gemstone Transaction Components

DESIGNERS
James Arakaki, Mark Stanton, Jeff Sasaki, Allan Avnet of VeriFone Inc.

CLIENT
VeriFone Inc.

System 3000 Models 3225/3230/3333

DESIGNERS
Werner Stephan of NCR Germany;
Graham Marshall, Robert Kelley, Brian Jablonski, John Caldwell of NCR Corp.

CLIENT
NCR Corp.

Signature Capture Terminal

DESIGNERS
Diana Juratovac, Gregg Davis,
Peter Koloski of Design Central;
David Allgeier of NCR Corp.

CLIENT
NCR Corp.

nCUBE 2 Supercomputer

DESIGNERS
Hartmut Esslinger, Howell Hsiao of
frogdesign inc.

CLIENT
nCUBE

CM-5 Massively Parallel Super Computer

DESIGNERS
Danny Hillis, Don Moodie of
Thinking Machines Corp.;
Maya Lin; Marc Harrison of
Marc Harrison Associates

CLIENT
Thinking Machines Corp.

Digitizing Scanners

DESIGNERS
Bill Moggridge, Roberto Fraquelli,
Andy Deakin of IDEO Product Development;
Gaetan Spake, Joe Murray of OCÉ Graphics

CLIENT
OCÉ Graphics

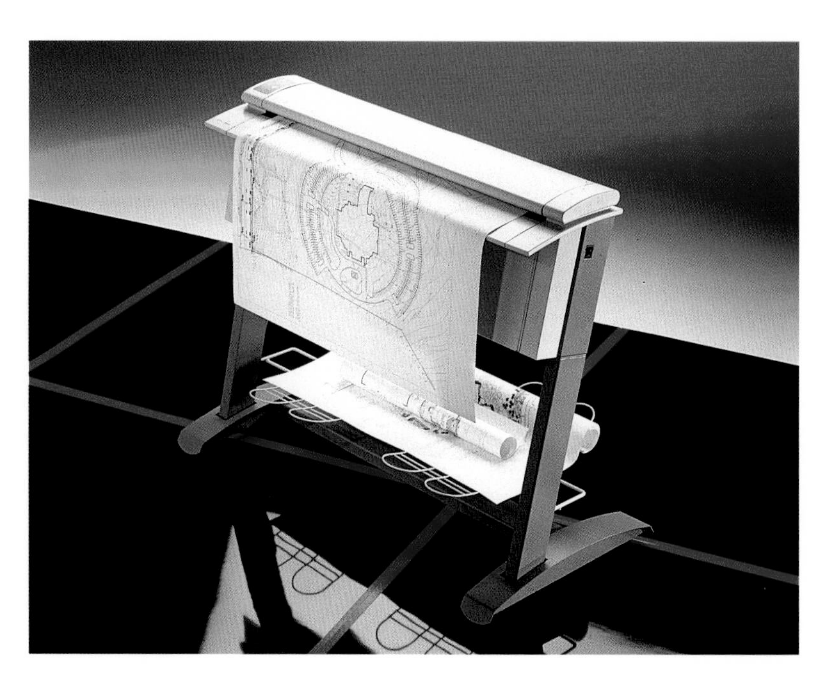

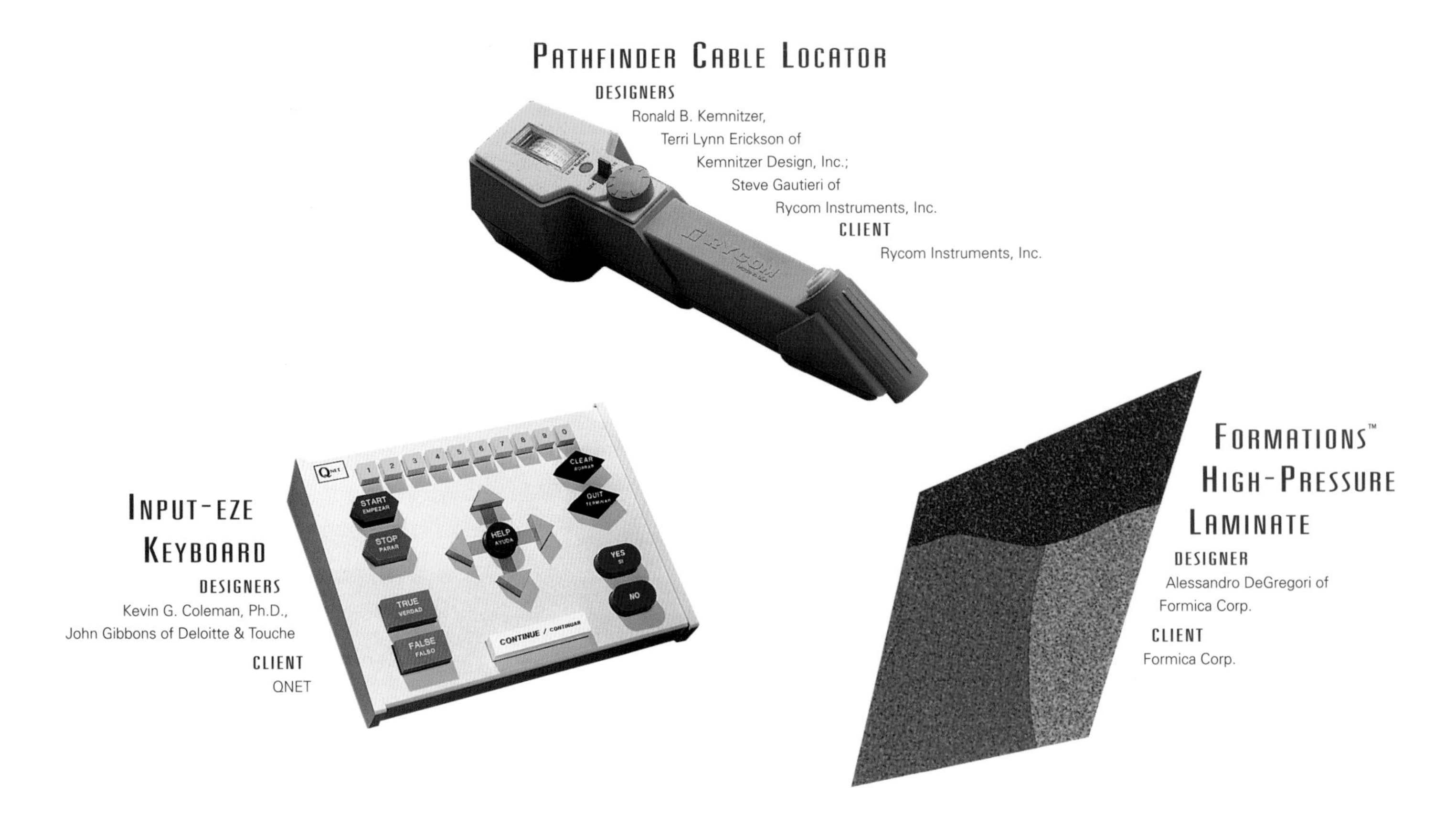

Pathfinder Cable Locator

DESIGNERS
Ronald B. Kemnitzer,
Terri Lynn Erickson of
Kemnitzer Design, Inc.;
Steve Gautieri of
Rycom Instruments, Inc.

CLIENT
Rycom Instruments, Inc.

Input-eze Keyboard

DESIGNERS
Kevin G. Coleman, Ph.D.,
John Gibbons of Deloitte & Touche

CLIENT
QNET

Formations™ High-Pressure Laminate

DESIGNER
Alessandro DeGregori of
Formica Corp.

CLIENT
Formica Corp.

Found-Wood Furniture System

DESIGNER
Frank J. Fusco

DESIGN SCHOOL
Pratt Institute

Phone Holder

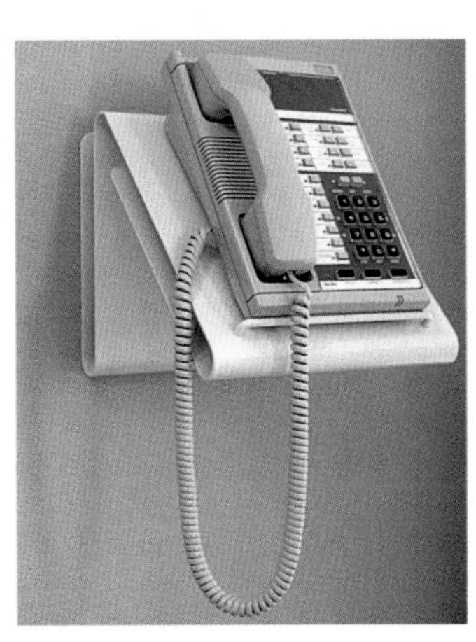

DESIGNERS
Dimitri Avdienko of
Skidmore, Owings & Merrill

CLIENT
American Honda Motors Co.

Working in the Office of the Future

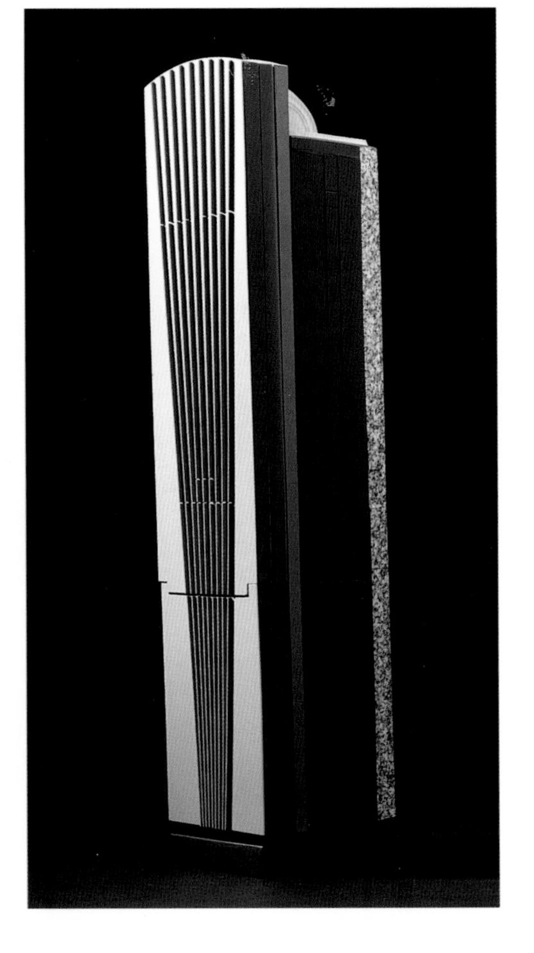

DESIGNER
Yohei Ota

DESIGN SCHOOL
Art Center College of Design

SPONSOR
Steelcase

Ventricular Assist System

DESIGNERS
Naoto Fukasawa, Tim Parsey, Robin Sarre, Jane Fulton of ID TWO

CLIENT
Baxter Healthcare Corp., Novacor Div.

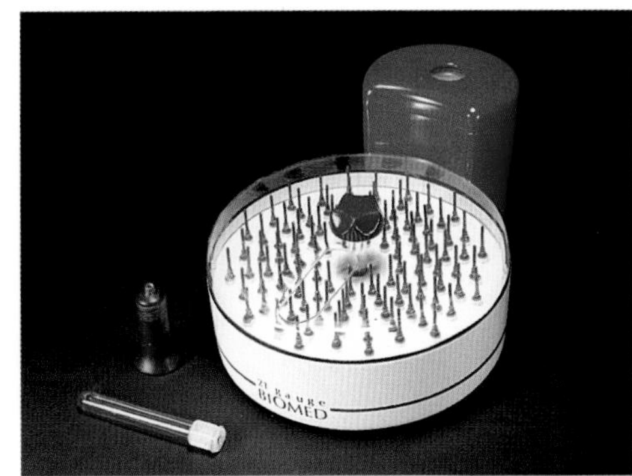

Needle Carousel

DESIGNER
Randy Bernard

DESIGN SCHOOL
Auburn University

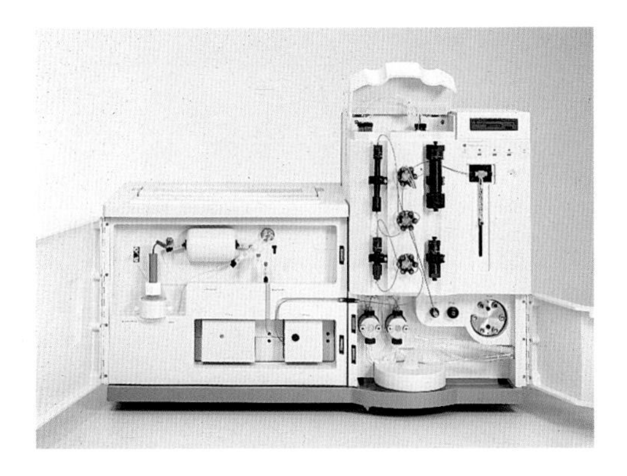

BioCAD Perfusion Chromatography Workstation

DESIGNERS
Luis Pedraza, Jose Tadeo de Castro of Design Continuum Inc.;
Neil F. Gordon, Ph.D. of PerSeptive BioSystems

CLIENT
PerSeptive BioSystems

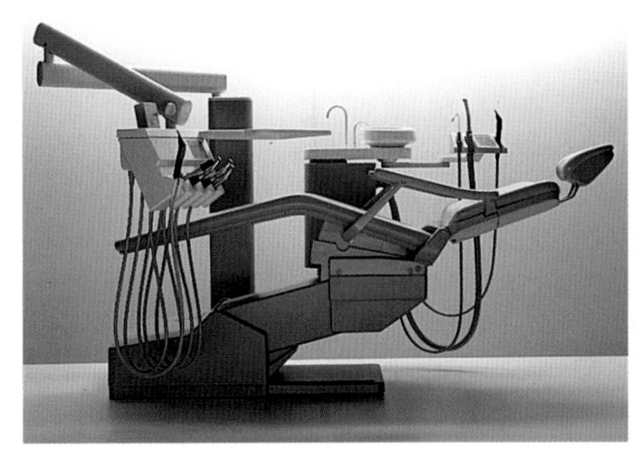

1062 Dental System

DESIGNER
the frogdesign team

CLIENT
KaVo GmbH

Split Second® Central Venous Catheter

DESIGNER
Timothy J. Erskine of Becton Dickinson Vascular Access

CLIENT
Becton Dickinson Vascular Access

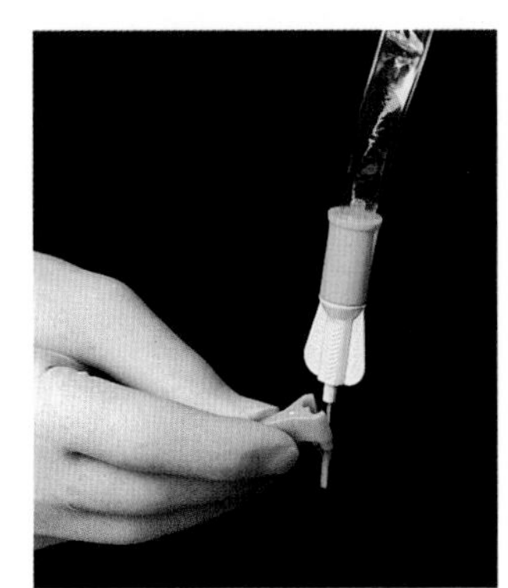

Anatomy Immersion Dissecting Table

DESIGNERS
John McBrayer of McBrayer Industrial Design;
Bill Joiner, Bernard Daniels, David Garza of KLN Steel Products, Co.

CLIENT
KLN Steel Products, Co.

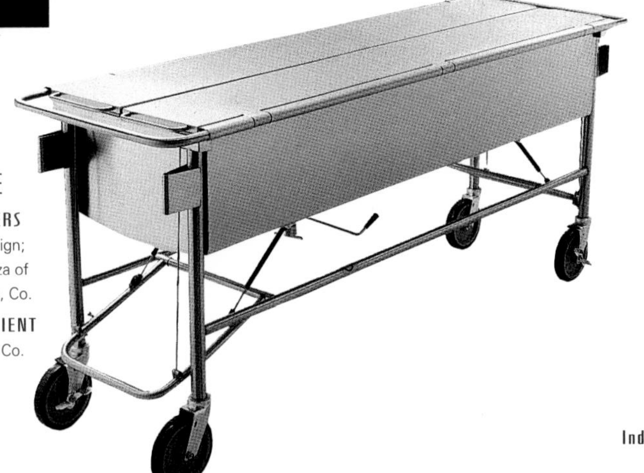

ACL DST-800

DESIGNERS
Ron Boeder of Boeder Design;
Darrell S. Staley of Ampex Corp.

CLIENT
Ampex Corp.

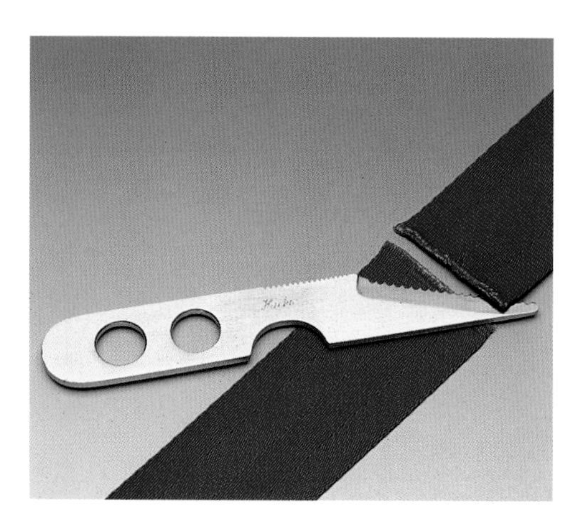

EMT-1

DESIGNERS
Gregory J. Hicks,
Jeff Hanna of u r o designs

CLIENT
Crawford Knives

DNA Sequencer

DESIGNERS
Bill Evans, Gil Wong,
Keith Willows, Marieke van Wijnen,
Max Yoshimoto of Lunar Design

CLIENT
Li-Cor, Inc.

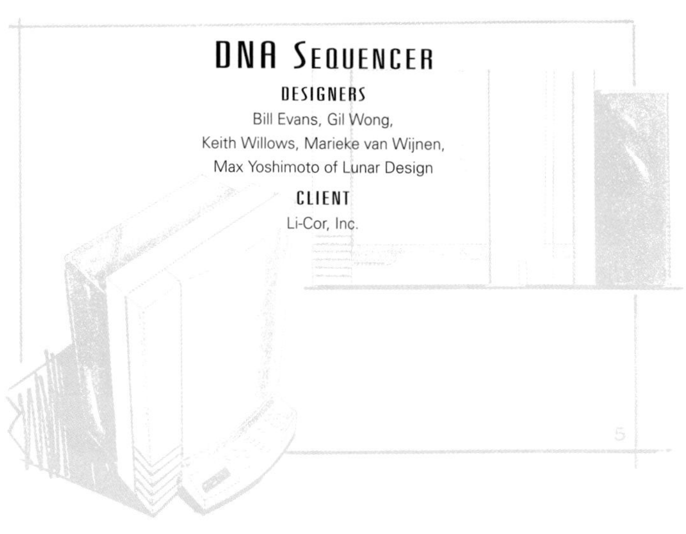

Imagine

DESIGNERS
Robert LaRoche, Martin Chenette,
Jean Barbeau, Alicia Starr of
Precision Mfg. Inc.

CLIENT
Precision Mfg. Inc.

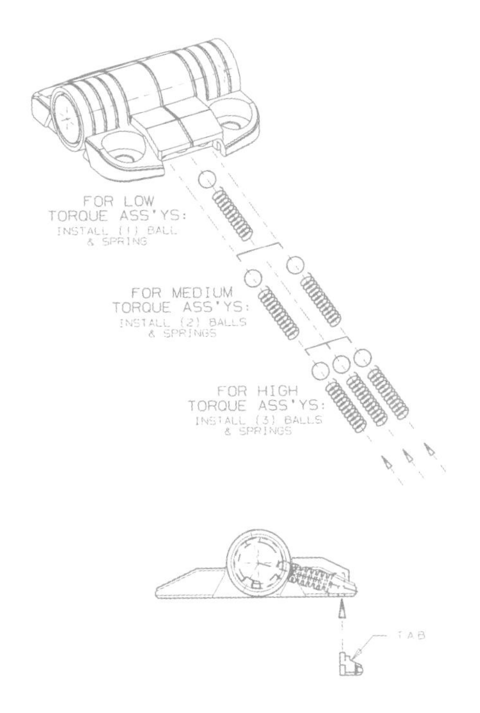

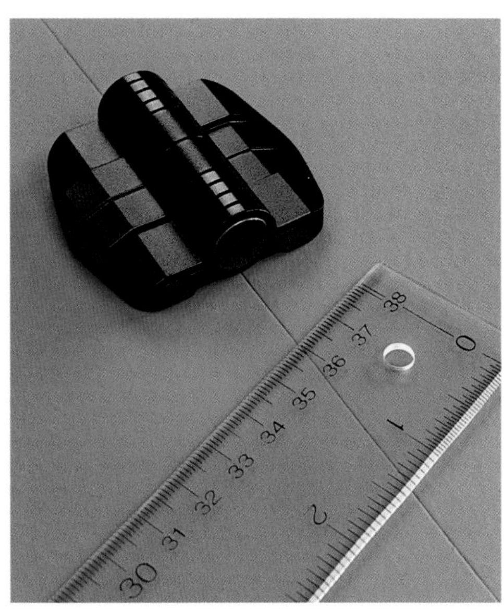
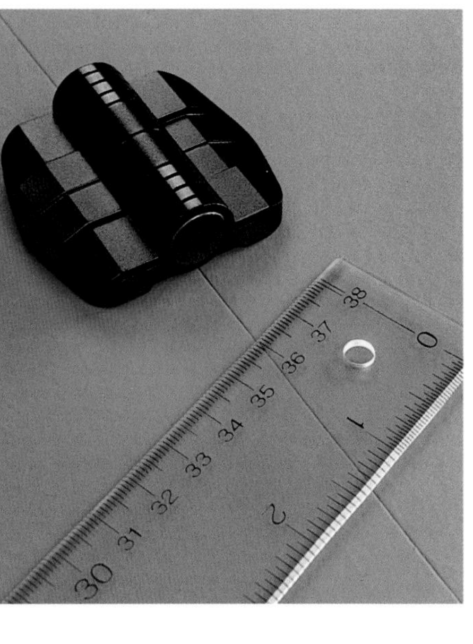

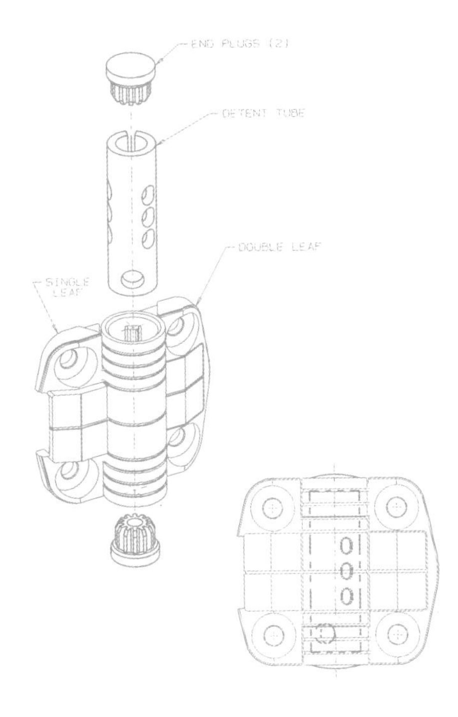

Door Positioning Hinge

DESIGNERS

Allen Riblett of Southco, Inc.;
Vince Juliana of Juliana Design Group, Inc.

CLIENT

Southco, Inc.

Thermedics Portable Explosives Detector

DESIGNERS

Michel Arney, Lynn Noble of
Design Continuum Inc.

CLIENT

Thermedics Inc.

Telecommunication Console

DESIGNERS

Corporate Design Group Industrial Design of
Bell Northern Research

CLIENT

Northern Telecom

Business Class Headphones

DESIGNERS

Sander J. Sinot,
Steven van Dijk of BPV/SDA

CLIENT

KLM Catering Services

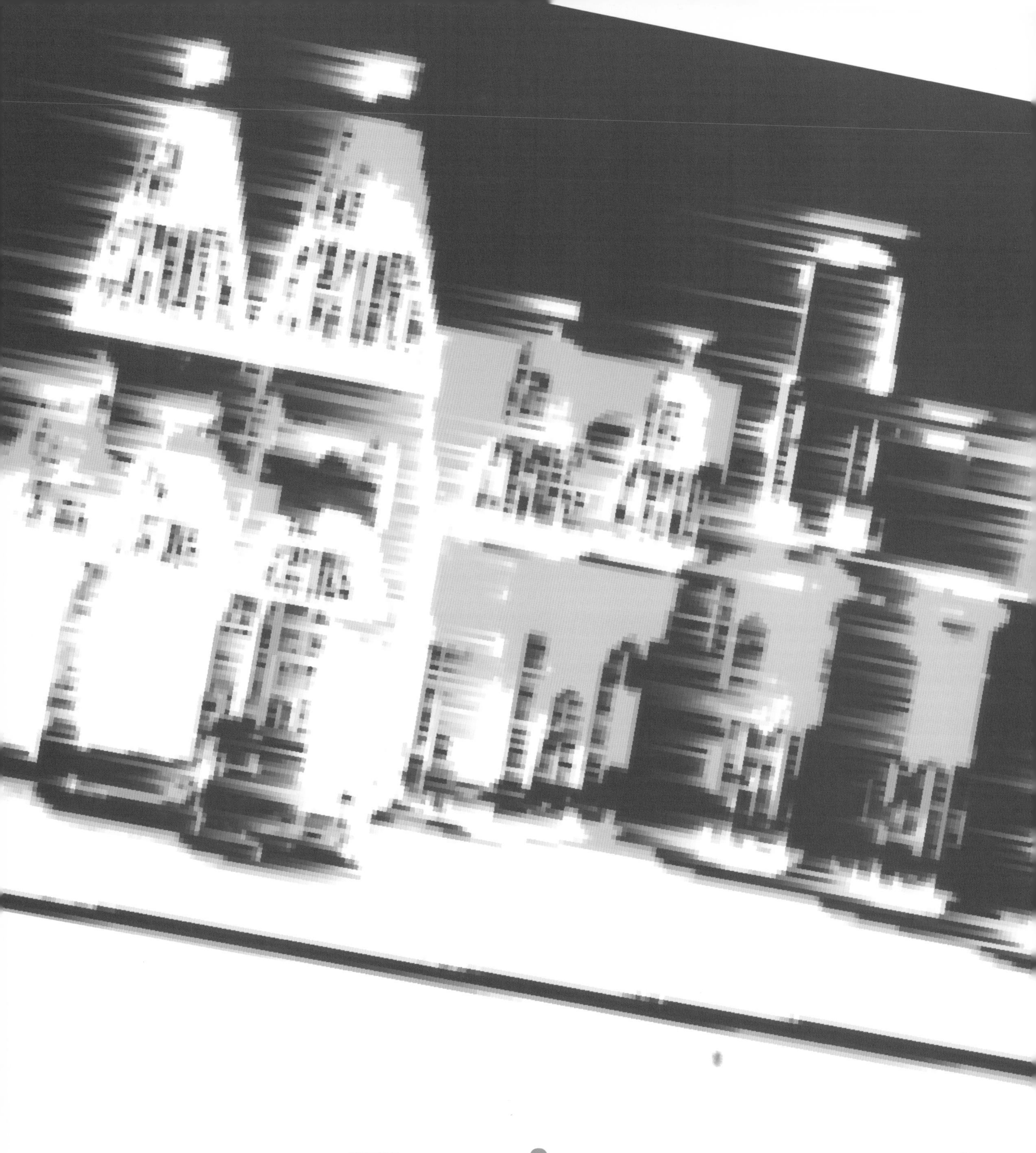

Environmenta
& Packaging
design

Lee Trade Show Exhibit Booths

Design Objective

The redesign needed to communicate the company's repositioning and its new role as a market-facing innovator in the casual apparel business. The booth's message had to be clear and consistent: Lee understands what the consumer wants and provides a product line that reflects that understanding.

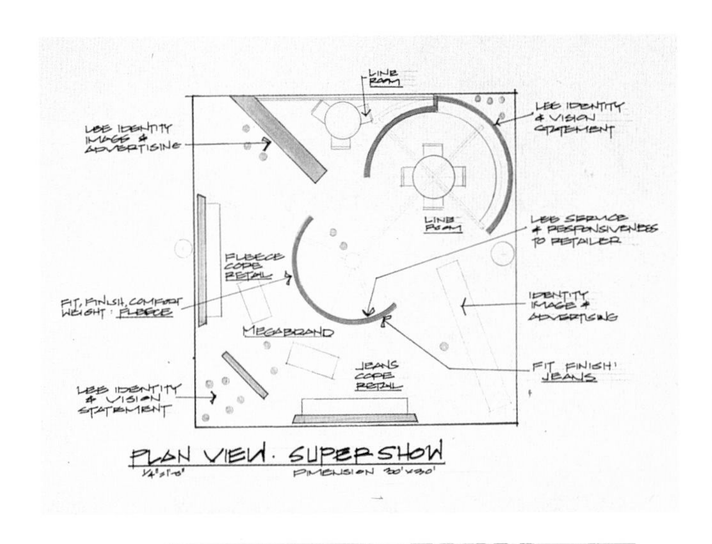

Design Solutions

Fitch Inc. emphasized lifestyle in designing the booth's overall effect, which successfully integrated the Bassett-Walker Co. knitwear with the Lee products and marketing approach. Wood, leather, sisal, and painted metal combined with "red slate" laminate and back-lit fiberglass enhance the new product assortment and reinforce Lee's down-to-earth positioning.

Result

This exhibit booth presents a new image of Lee to customers, introducing the new integrated products—sweats plus jeans—while presenting the products the way Lee wants its retailers to present them.

Designers

Jaimie Alexander, Paul Lechleiter, Matthew Hern, Paul Westrick, Gabriel Murray, Mindie Trank, Sandy McKissick, Carole Kincaid, Neil Powell of Fitch Inc.

Client

Lee Company

"Through the use of vernacular imagery in the architectural shell of the booth, the designers signed the most vernacular of American clothing—denim. The environment and fixtures were solidly executed through strong detailing and a good variation in scale and material finishes...."

JUROR AURA OSLAPAS, IDSA

"This project was successful in that it helped to communicate a brand and business strategy for Lee. It combined visual impressions with strong verbal messages to clearly define Lee's understanding of the consumer and of the trade."

FITCH INC.

From White to Green

"...Apple Computer has made an arresting design statement here while maintaining the superb design, protection and communication qualities of the original packaging...this project highlights the fact that greening doesn't mean increasing cost or giving up quality—real or perceived."

JUROR RITASUE SIEGEL, IDSA

DESIGNERS
Peter T. Allen, Jean Stevens, Charlie Costantini, Peggy Jensen, Greta Mikkelsen, Chantale Hansen of Apple Computer, Inc.

CLIENT
Apple Computer, Inc.

DESIGN OBJECTIVE

To revitalize Apple's packaging design, eliminate the use of bleached white packaging materials by using recycled materials instead, and reduce both the material and production costs of Apple's packaging.

DESIGN SOLUTIONS

Apple dispensed with the bleaching process, one of the more polluting aspects of paper and cardboard preparation, and selected less toxic inks and printing methods: water-based flexo inks without heavy metal contaminants.

RESULT

The redesign is bolder, more confident, and signals Apple's continued commitment to improvement. The new packaging resulted in an annual savings of $3.2 million in the first year and a 50 percent materials cost reduction.

La Croix Water Packaging Redesign

DESIGN OBJECTIVE

To create a new identity for La Croix that would appeal to younger consumers, stand out on a crowded shelf, and communicate an image of refreshment and healthfulness.

DESIGN SOLUTIONS

The design team spent two weeks at the can company's facilities using a proofing press and its inking department in order to achieve the graduated color on the uncoated aluminum can. Cleanly presented typography combined with a minimum amount of graphics customizes the story of La Croix.

RESULT

Despite the return of Perrier to the shelves, this package has maintained double the sales of its predecessor, with no increase in advertising or other promotional efforts. The new package is also credited with helping the client sell the product into all major airlines' in-flight services, increasing brand sampling.

"The La Croix packaging redesign works because, upon first impression, it looks and feels like what it is—fresh, cool sparkling water,"

JUROR DAVID FREJ

DESIGNERS

Tony Lane, Tom Decker, Paul Coyne of The Design Company

CLIENT

La Croix Bottled Water Company

The Garage

DESIGN OBJECTIVE

To design an exhibit that would make technology accessible to the general public and interesting as a career choice for young people.

DESIGN SOLUTIONS

Rafter-like beams in the lobby help direct traffic and break up what would otherwise remain a monotonous thoroughfare. Each exhibit area includes a curved white "textbook" wall of background information which lays the groundwork for the second set of interactive exhibits. With the exception of painted "textbook" walls, all other materials retain their natural finishes and are only protected with coating when necessary.

RESULT

The exhibit is a warm, casual and quiet environment with dense, layered and multi-sensorial learning experiences. The exhibit has been well-received by the general public as well as by experts in the various technologies presented.

OTHER AWARDS

–*I.D.* Travelers Bureau,
One of the Top Ten Attractions of 1990

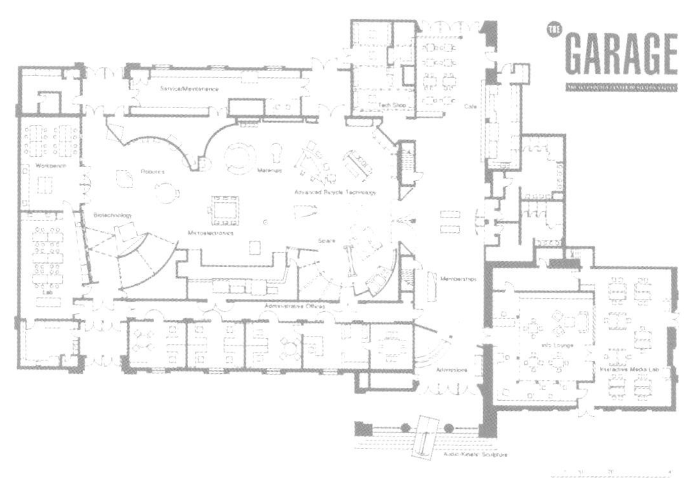

"The designers have organized lots of information beautifully in this exhibit. The jury especially admired how this design is not the least bit glitzy, all but disappearing and leaving the information uncluttered by distracting input."

JUROR DIANNE PILGRIM

DESIGNERS
Aura Oslapas, Pirkko Lucchesi, Maureen Seitz, Steven Tornallyay, Ken Krayer, Phred Starkweather, Tim Power of A + O Studio; Karl Martens

CLIENT
The Technology Center of Silicon Valley

Pill Dispenser Package

"The challenge for the designer was to communicate what the product is and why you should own it. This package beautifully communicates those issues to the consumer. There is no question about the purpose with its clear, direct graphics."

JUROR NOEL MAYO, IDSA

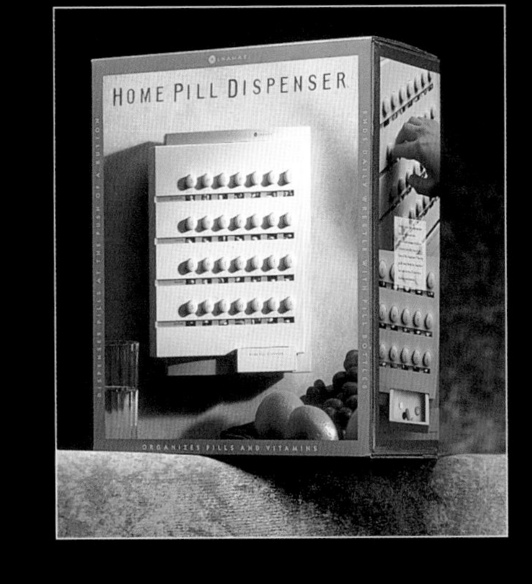

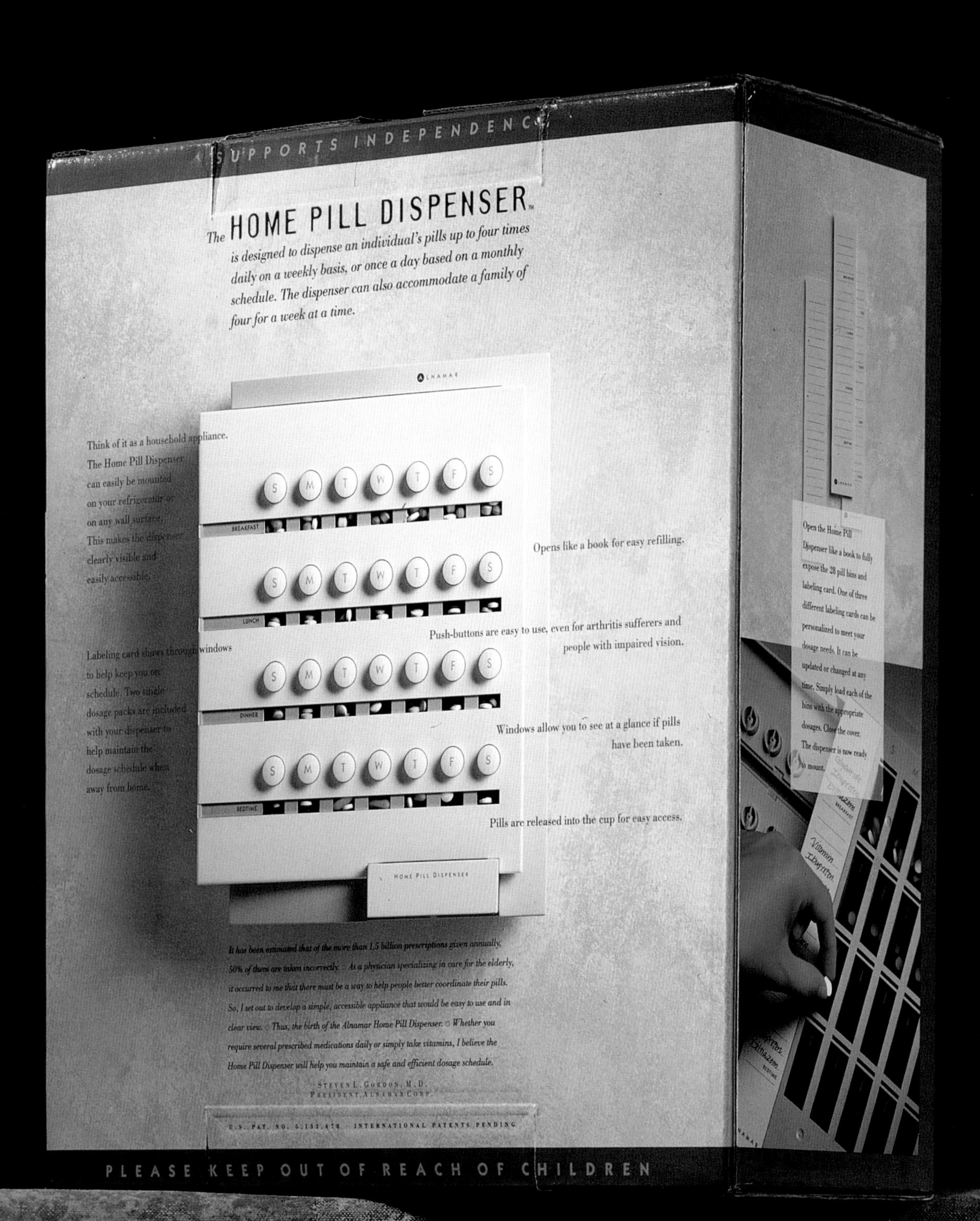

DESIGN OBJECTIVE

To develop a graphic design for a package that would help sell the Alnamar Pill Dispenser at retail. The package needed to function as a shipping container, a point of purchase display, and a reusable carrying case for the customer.

DESIGN SOLUTIONS

The package graphics anticipate and lead the customer's education process. Rich earth tones and a muted yellow background on the face photography draw customers to the package with a powerful billboard effect. The copy on the side panels "pulls" the reader to the back of the package.

RESULT

This packaging design is both traditional and elegant, appealing to older consumers. Within 90 days of package introduction, the number of active retail accounts increased from 55 to more than 275, and sales increased by more than 300 percent.

DESIGNERS
Don Rood, Sohrab Vossoughi of ZIBA Design

CLIENT
Alnamar Corp.

Learning & Learning Disabilities: Explorations of the Human Brain

DESIGNERS
Gregg Loeser, Anne Hornickel, Grant Haring of the Museum of Science & Industry;
Dirk Wales of Rainbow Productions;
Jane and Ed Bedno of Bedno/Bedno

CLIENTS
Museum of Science & Industry;
The J. Ira Harris Foundation; The Pritzker Foundation

DESIGN OBJECTIVE

To develop an exhibit about brain function and learning disabilities that would allow public access to some of the known complexities of the human brain.

DESIGN SOLUTIONS

The exhibit was organized to include an introduction, physiology learning center (information process, brain dysfunction), experience pods (what a learning disability is like), a real-people video presentation, and a resource center. The activities of a birthday party became the common thread which linked all of the content information. Giant brains floating above giant ½ heads create a central architectural presence.

RESULT

The dramatic appearance of the design vocabulary is visible from a distance, and draws the visitor to the subject. The design vehicle encourages periodic content update to keep the exhibit focus on the cutting edge. Additionally, it was the first exhibit at the museum to offer research reference to continuing education as an integral part of the learning experience.

OTHER AWARDS

–American Association of Museums,
Highest Recognition for Excellence in Exhibit Content and Design Concept

"...The TVs are an effective device that help to give focus to the exhibit. Although there's sensory assault in this exhibit, it works."

JUROR DIANNE PILGRIM

"A successful exhibit design creates immediate emotional visitor access by being a distinct time and place that expresses the inner logic of the exhibit content. Once it has captured the visitor's subconscious, it accesses that visitor intellectually by presenting its content at different levels and by different methods of communication chosen to enhance each particular conversation."

GREGG LOESER

Birmingham Civil Rights Institute Exhibit

DESIGN OBJECTIVE

A good exhibition must attract, hold interest, and communicate. The exhibit was designed to commemorate the role of Birmingham in the civil rights movement and look into the future to an improved understanding of civil and human rights.

DESIGN SOLUTIONS

This exhibit communicates the historical context, social setting, struggles and triumphs of the civil rights movement, focusing on the era between 1956 and 1965. The design solution presents the information in a linear, chronological story line. The most innovative aspect lies in the variety of presentation modes and skillful integration of such media as documentary film, interactive video, two forms of video walls, multi-image slides and archival footage. Text was broken into 50-word "chunks" to increase readability.

RESULT

After an introductory film, visitors travel through five separate galleries, each sequencing a specific time period. Entitled "Barriers," "Confrontation," "The Movement," "The Processional," and "Milestones," each gallery transports the visitor to that particular time period, allowing him or her to feel the pain as well as the triumphs.

DESIGNERS

Howard Litwak, Gail Ringel, Sherry Proctor, Chris Danemayer, Jennifer Sargent of Joseph A. Wetzel Assoc., Inc.

CLIENT

The City of Birmingham, Alabama

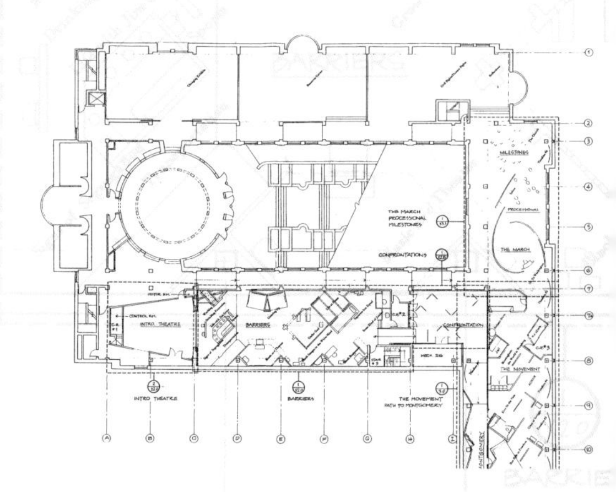

"I commend the City of Birmingham for implementing this meaningful and significant exhibit. Its use of period film on the '60s television sets, combined with the colorless cast of figures, captures those historic moments perfectly...."

JUROR NOEL MAYO, IDSA

"...Based on the hundreds of visitor responses we obtained, it is obvious that the exhibit galleries are very effective for transporting people to the time and place described in the exhibits."

DR. STEPHEN BITGOOD OF THE CENTER FOR SOCIAL DESIGN IN ANNISTON, ALABAMA

Individual Trainer Merchandising System

DESIGNERS
Anton Kimball, Steve Wittenbrock of
Wittenbrock Design

CLIENT
Leap Inc.

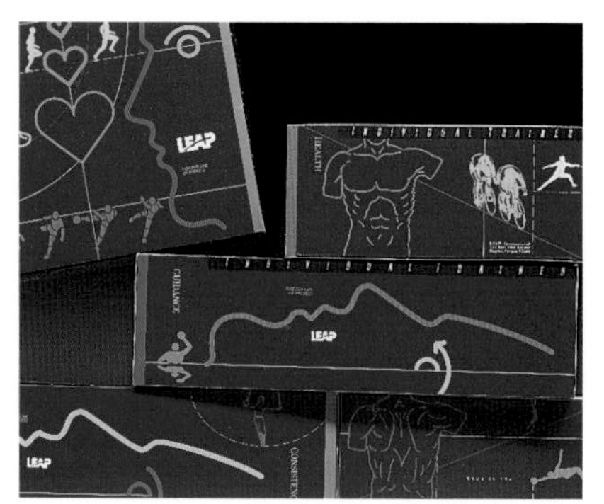

The Individual Trainer is a computer that enables people to develop workout programs and monitor their recreational activities. The wraparound graphics of this system show possible uses of the computer with images of vitality, sport and humor. The multipaneled approach to the layout implies motion, encourages hands-on inspection, and adds to shelf impact.

Airplane Passenger Cabin Interior

DESIGNERS
777 Teague Design Group;
Alan Mulally, John Roundhill,
George Broady, Duncan Mullholland,
Art Pompei, Jim Huentleman of
Boeing Commercial Airplane Group, 777 Division

CLIENT
Boeing Commercial Airplane Group

This design program creates an environment that is spacious and flexible. Market driven, it results from formal research. The curvilinear architecture is visually more pleasing. Retracting storage bins increase the physical and perceived space.

Recyclery Education Center

DESIGNERS
Andrew Kramer, Tim Kobe, Greg Kono,
Bill Chiaravelle of West Office Design Associates, Inc.

CLIENT
Browning-Ferris Industries, Inc.

The Recyclery is an educational exhibit at an integrated waste management facility. A 16-foot-high Wall of Garbage runs the full 100 feet of the exhibit, showing the amount of trash generated in our country in one second. Other exhibits let visitors sort cans with an electromagnet, tour the landfill with an interactive video program and follow the recycling steps of glass, aluminum, paper and plastics.

PAYFAX International Symbol

DESIGNER
Dave Nicol of IDE Inc.

CLIENT
FAX Plus Inc.

This symbol was designed to communicate visually to people of all nationalities that there is a public fax facility available for use nearby. By linking three standard images, the design solution visually conveys the idea of transmitting graphic documents digitally over telecommunication lines.

Mouseman Packaging

DESIGNERS
Timothy Stebbing, Beverly Catli
of frogdesign inc.

CLIENT
Logitech Inc.

The package emphasizes the ergonomic and curvaceous form of the mouse and the mouse's right- and left-handedness. The information is clearly displayed without undue clutter or surrender of spirit.

Portable Privacy Station

DESIGNERS
Roy Fischer, Randall Toltzman of Designology Inc.;
Phillip Kolo, Michael Lamb,
Gary Pinkstaff of Blood Systems Inc.

CLIENT
Blood Systems Inc.

The Food and Drug Administration (FDA), which regulates procedures for the collection of donor blood, strongly recommended that a private environment be used during the interview which prequalifies blood donors. The Portable Privacy Station provides that privacy quite simply: it is a 5- by 19-foot, one-sided corrugated, polyethylene sheet, stabilized on each end with a split PVC tube and fabricated base. Weighing 38 pounds, it rolls up easily for transport.

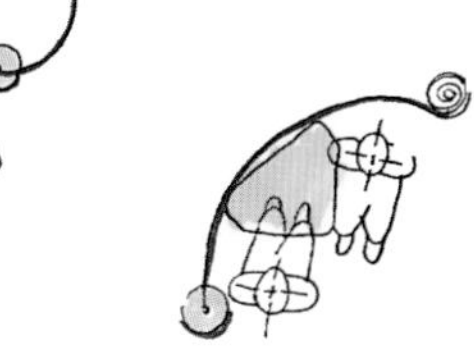

The Power of Maps Exhibition

DESIGNERS
James Biber,
Peter Harrison of Pentagram

CLIENT
The Cooper-Hewitt Museum

About the significance of maps as instruments of communication, persuasion and control, *The Power of Maps* exhibit involves visitors in its exploration. Surveys of visitors prior to viewing the exhibition and upon exiting it showed a significant shift in attitudes, indicating the exhibition was successful.

M Series Packaging

DESIGNERS
Dan Ashcraft, Terry Scott,
Sue Slutsky of Ashcraft Design

CLIENT
JBL International

This packaging for automotive amplifiers abstracts the car amplifier in photography so as to make details of the amplifier into artistic pieces in their own right. It organizes the information in such a way that consumers can quickly understand the new product's features and realize more fully exactly what is being purchased. The entire package is printed and laminated to recycled papers.

GT Series Automotive Loudspeaker & Amplifier Packaging

DESIGNERS
Dan Ashcraft, Terry Scott,
Judi Weber of Ashcraft Design

CLIENT
JBL Consumer Products, Inc.

These packages for automotive loudspeakers and amplifiers, made of brown kraft, are used for both shipping and display purposes. An innovative two-color design reduced packaging costs by 40 percent. Colors and images reflect the high energy, power and motion associated with sports for the 18- to 24-year-old-male target audience.

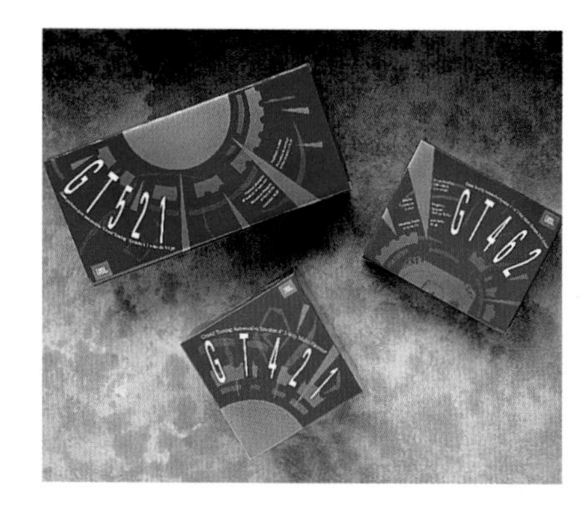

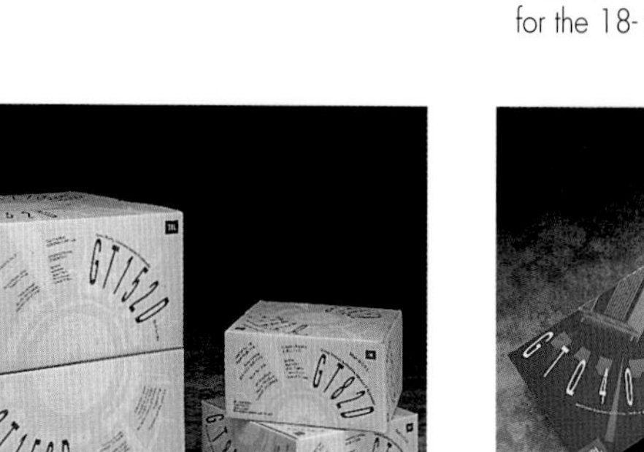

Thirty Under Thirty

DESIGNERS
Thomas Gamble, III, Kieth Mascheroni for Steelcase Design Partnership

CLIENT
Steelcase Design Partnership

This exhibit provides a forum for the ideas of the up-and-coming generation of designers from all over the world. Steel poles and cables standing in a painted sonotube base support photographic representations. This structure makes the exhibit both lightweight and easy to transport, requiring no skilled professionals for installation and breakdown.

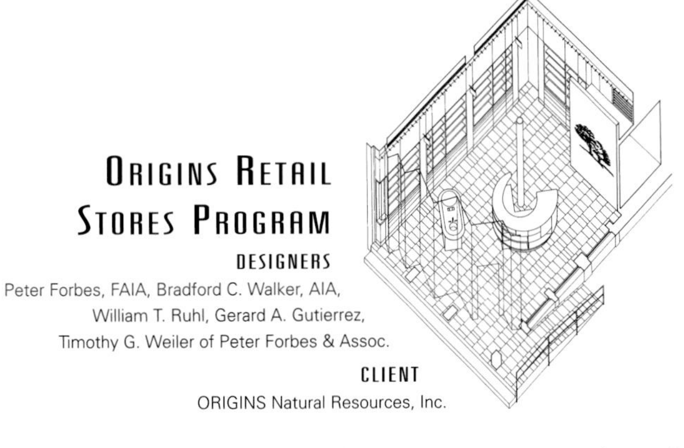

Origins Retail Stores Program

DESIGNERS
Peter Forbes, FAIA, Bradford C. Walker, AIA, William T. Ruhl, Gerard A. Gutierrez, Timothy G. Weiler of Peter Forbes & Assoc.

CLIENT
ORIGINS Natural Resources, Inc.

This store design reflects the company's commitment to environmentalism and Origin's respect for its customers through each store's choice of materials, recycling efforts, attention to air quality and relation to the surrounding neighborhood. The layout of the design emphasizes a removal of barriers allowing the customer to approach and sample the products.

Frankfurt Furniture Showroom

DESIGNERS
Erik Sueberkrop, Peter VanDine, Arthur Collins, Brigit Fichtner of STUDIOS Architecture, Paris; Max Durney of Limited Productions, Inc.

CLIENT
The Knoll Group

This showroom reflects Knoll's heritage and tradition while projecting a progressive image. Visitors follow the process of raw materials through design and engineering to the finished product. Knoll's products are used throughout—in the reception area, the manager's office and conference space—allowing customers to see the furniture in use.

Furniture Product Brochure

DESIGNERS
Janet P. Rauscher, Stephen Witte of Rauscher Design Inc.; Image Source Photography

CLIENT
Versteel

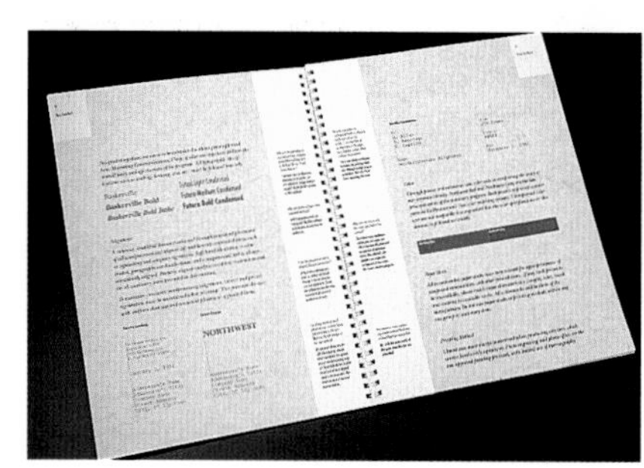

Stationery Guidelines

DESIGNERS
Don Kline, Charles Spilman of Landor Associates

CLIENT
Northwest Airlines

SignTrack™

DESIGNERS
Ron Cobb, Ben Bell of APCO

Graphika

DESIGNERS
Carol Bouyoucos, John Waters of Waters Design Associates

CLIENT
James River Corp.

Emilio Ambasz Exhibit Poster

DESIGNER
Emilio Ambasz of Emilio Ambasz Design Group

CLIENT
San Diego Museum of Contemporary Art

Ultradata Logo

DESIGNERS
Martha Davis, Lisa Krohn of Able

CLIENT
Ultradata

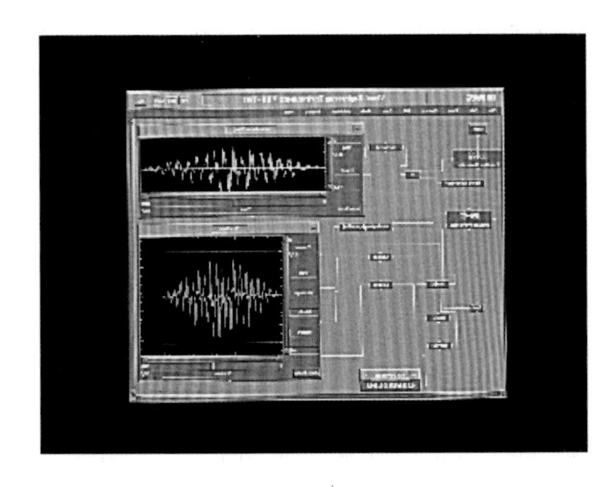

Visual Engineering Environment

DESIGNERS

Jon K. Pennington, Bill Hunt,
John Bidwell of Hewlett-Packard

CLIENT

Hewlett-Packard, Measurement Software Division

User Interface for System Manager

DESIGNERS

Peter Spreenberg, William Verplank of
IDEO Product Development;
Susan Wolfe, Brady Farrand,
Roland Findlay of Tandem Computers, Inc.

CLIENT

Tandem Computers, Inc.

FroxSystem User Interface

DESIGNERS

Curtis Abbott, Herb Jellinek, Dan Freed,
Randy Wiggington, Gareth Loy of Frox, Inc.;
Marsh Chamberain of Axiom User Interface Ltd.

CLIENT

Frox, Inc.

Businessland/ComputerCraft Prototype Store

DESIGNERS

Bruce Burdick, Susan K. Burdick, Bruce Lightbody,
Cindy Steinberg of The Burdick Group

CLIENT

ComputerCraft

Park City Center Banners

DESIGNERS
MB Flanders, Arthur Ganson,
Polly Baldwin of Flanders + Assoc.;
Flagraphics

CLIENT
Dusco Property Management

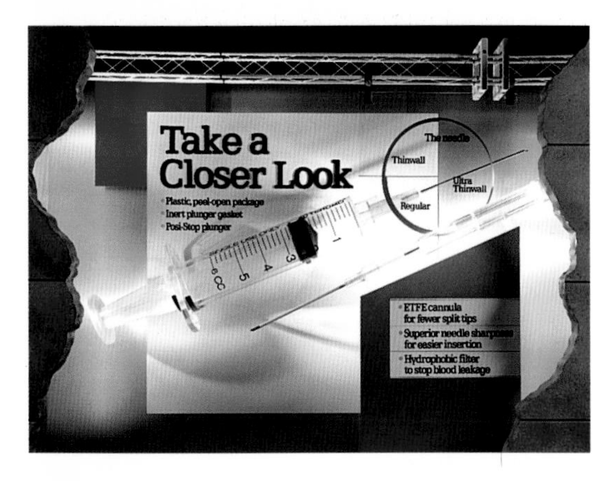

Terumo Corporate Exhibit Program

DESIGNERS
Jack Snyder, Tim Dexter, Gregory Meadows,
Andrea Billias of Design for Industry

CLIENT
Terumo Corp.

Display at Orgatec

DESIGNERS
Erik Sueberkrop, Peter VanDine,
Brigit Fichtner, Andre Straja of
STUDIOS Architecture, Paris;
Roland Rudolph of 3R

CLIENT
The Knoll Group

Tucson Mall Teapot Pylon

DESIGNER
Suzanne Redmond Schwartz of
RTKL Associates Inc.

CLIENT
Forest City Enterprises

Acknowledgments

The Industrial Designers Society of America would like to acknowledge Business Week *magazine and its design editor, Bruce Nussbaum, for their role as a key innovator in US business.*

Jury Chair

Gerald P. Hirshberg, IDSA

Jurors

Charles H. Burnette, Ph.D., IDSA
David Frej
Dianne H. Pilgrim
William Stumpf, IDSA
Tucker Viemeister, IDSA
Arnold Wasserman, FIDSA

WINNERS

7785 Service Bay System • Access *(concept)* • Access Power•Pointer *(concept)* • Animal • Blood Banking System Research • The Brick • Bunjieboarding—Bunjie System *(concept)* • Concept 2000 Computer System *(concept)* • Dodge Neon *(concept)* • Drake-Willock™ System 1000 Dialysis Machine • ECO35 • The Garage • JuiceMate *(concept)* • La Croix Sparkling Water Packaging • Large Screen LCD Video Projector *(concept)* • Learning & Learning Disabilities: Explorations of the Human Brain • Medical Ultrasound Imaging System • Suburban Public Phone *(concept)* • Tadpole Infant Positioning System • Work Group DEChub, DECserver & DECbridge

Silver

WINNERS

ACS Angioplastic Catheter • AddressWriter • Attiva Seating System • Busse Bac/shield • Contour *(concept)* • DataVault & Connection Machine-2A • Deni Freshlock-Vacuum Sealer • Digital Answering System 1337 • Encore • Ergonomic Correction Pen • FiberWorld • Flexor® Gloves • "Goldilocks and the Three Bears" *(concept)* • GRiD 1810 • GRiDPAD • Hand-Tite™ Keyless Chuck • High Density Disk Array • Individual Trainer Merchandising System • International Design and Engineering Accessment Truck *(concept)* • Long Reach Flexible Flashlight • M Series Packaging • Microphone and RCA Adapter • MOTO 1 Motorcycle *(concept)* • Mouseman Packaging • Oseris 12V/50W Spotlight • PAYFAX International Symbol • Point of Sale Graphics Research • Portable Presentation System *(concept)* • Prizm Video Workstation • Sand Soc Footwear • Series FC Four-Wheel Counterbalanced Sitdown Truck • Spacecab Medicine Cabinet • Spoons to Grow With • Stingray Trackball • Sushi Phone *(concept)* • SwingArm Typing Aid *(concept)* • Telesketch *(concept)* • Telesketch *(concept)* • Thirty Under Thirty • Trilliant™ Home Power System • Woodzig Power Pruner • Zero *(concept)*

WINNERS

AccuPinch • Backpacking Tents • BallPoint Mouse • BCN • Bulldog Chair • Businessland/ComputerCraft Prototype Store • Component Tree *(concept)* • Corded Mouseman • Cyclone Grinders • Emilio Ambasz Exhibit Poster • The Eternal Frame *(concept)* • F20706FT 20-inch Television • Flash Tracks • Fluorescent Lighting Products • Formations™ High-Pressure Laminate • Found-Wood Furniture System *(concept)* • Fun Hydrant Sprinkler • Furniture Product Brochure • Garment Bag Plus • Gemstone Transaction Components • Graphika • Handi-Scratch Wire Brush • HEARSAY *(concept)* • Helper *(concept)* • Individual Microwave Drip Coffeemaker Concept *(concept)* • Input-eze Keyboard • Lincoln *(concept)* • Millipore Third World Environmental Incubator • nCUBE 2 Supercomputer • Orbital Finishing Sander *(concept)* • Park City Center Banners • Phone Holder • Portable Computer Product Family • Potty Seat • PS/1 Computer • Qualtronics Fax Machine • Sawtch Reciprocating Saw *(concept)* • Smoke Evacuator • St. Tropez • Stationery Guidelines • StyleWriter Inkjet Printer • Tecno Qualis • Terumo Corporate Exhibit Program • Thermedics Portable Explosives Detector • Trigem 386SX Laptop Computer • Tucker Recycling System • Tucson Mall Teapot Pylon • Typist Scanner • Ultrdata Logo • Ventricular Assist System • Working in the Office of the Future *(concept)* • YST-SD90 Audio Speaker • Zwirl

1992

JURY CHAIR
Arnold Wasserman, IDSA

JURY
James Bleck, IDSA
Stephen Hauser, IDSA
David Jenkins, IDSA
Fritz Mayhew, IDSA
Aura Oslapas, IDSA
Liz Powell, IDSA
RitaSue Siegel, IDSA

Gold WINNERS

AirBass *(concept)* • Airflex • Aquatread • BeeperKid • CleanWorks • CLINITRON® ELEXIS™ Air Fluidized Therapy • Demon Dispenser • "D" Series Angle Wrenches • Electrical Vehicle Charging System *(concept)* • EZ Router™ • Fit • From White to Green • Good Grips Kitchen Utensils • Handkerchief TV *(concept)* • Lee Trade Show Exhibit Booths • New Move Wheelchair *(concept)* • Orchestra Lamp • PowerBook • Precedence Bath • Proformix Keyboarding System • Rear Projection TV • Relay Furniture • Seville STS • Ultralite *(concept)*

Silver WINNERS

1-2-3 Bike • 30 Series Clampmeters • 3125 NotePad • 5775 Digital Color Copier • Airplane Passenger Cabin Interior • Attaché Portable Computer *(concept)* • BodyBoat *(concept)* • Cachet Articulated Stamp • Collection Fleury Stacking Chair • Crime Shield Window Barriers • Direct Manual Braille Slate • Dive-Alert • Finishing Planer *(concept)* • Flexible Endoscope • GunSafe *(concept)* • Image RV Awning System • In Search of an Unsurpassed Reputation for Design Excellence • Io Positional Track Lighting *(concept)* • Ithaca Solid-Ink Color Printer *(concept)* • Kangaroo® Pet™ Ambulatory Feeding System • Key Chain Security Remote Transmitter • Kidz Mouse • Metraloop • Metroblade • Mia Breast Pump *(concept)* • Momenta Pentop Computer • ON.IX • Plastic Combination Lock *(concept)* • Portable Privacy Station • Portions™ • Pressurized Lunar Rover Vehicle *(concept)* • Rage Jet Boat • Recyclery Education Center • Römer "frogdesign one" • Scale/Printer • Seatcase Compact Wheelchair • Series 300 IC Wood Sanding Machine • Showbox Photo Viewer • Spectrum NDT Transducer • Speedo® SwimMitt™ XT Aquatic Cross Training Gloves • Squejé® Aqua Tooth Polisher® • T-245 Metropolitan Toaster • Ultima 2000 Argon Laser System • 'vik-ter Stacking Chair

Bronze WINNERS

250T Them-A-Bind® • 1062 Dental System • ACL DST-800 • Agromat *(concept)* • Binoculars • BioCAD Perfusion Chromatography Workstation • Business Class Headphones • Catseye Candle Holder • CM-5 Massively Parallel Super Computer • Electric Moto Cross (EMX) • EMT-1 • FroxSystem User Inteface • Imagine • Infinity Radio Control • Jazzman CD Player *(concept)* • Jeep *(concept)* • Kool Mate 36 • Le Pet Cafe® • "Macintosh PowerBook vs. PCs" Product Value Analysis • Mirage Chaise Longue • Multi-scale Fretboard • Needle Carousel *(concept)* • Nomadic Workstation *(concept)* • PalmPAD • Pathfinder Cable Locator • PoqetPad • Portable Multimedia PC *(concept)* • Pure Air Ultrasonic Humidifier *(concept)* • ROLLiter • Saf-T-Loc • Salt and Pepper Shakers *(concept)* • Signature Capture Terminal • SmartLevel Series 200 • Soft Notebook Computer *(concept)* • Split Second® Central Venous Catheter • Telecommunication Console *(concept)* • Terrace • Toothbrushes • Trackman II • User Interface for System Manager • Visual Engineering Environment • Yontech 105

1993

Jury Chair

Peter W. Bressler, FIDSA

Jury

Charles L. Allen, IDSA
Vincent M. Foote, FIDSA
Louis E. Lenzi, IDSA
Noel Mayo
Ralph F. Osterhout, IDSA
Sharyn A. Thompson, IDSA
Chipp Walters, IDSA

Gold Winners

670 Perimeter • 3495 Tape Library Dataserver • Anthropometric Measuring System—"Anthropometron" • AtLite Exit Sign • Biojector • Birmingham Civil Rights Institute Exhibit • Deskjet Portable Printer • FirstPlay • FM 560 Motorized Stairclimber • Genesis Softside Easyturn™ Luggage • Highlander Concept Vehicle *(concept)* • imageRING *(concept)* • LAISer II • Leapfrog *(concept)* • LH Midsize Sedans • LiveBoard • Macintosh Color Classic • Metaform Personal Hygiene System *(concept)* • My First Sony Electronic Sketch Pad and Animation Computer • Northstar V-8 Engine • Palm-Mate • Peak 10 Camp Stove *(concept)* • Pill Dispenser Package • PowerBook Duo System • Retractable Wall Cord • Sensor for Women • Silhouette Window Shadings • Single Burner Portable Cooker *(concept)* • Softouch Scissors • Thermal Electric Grill • Tutor Training Table System

Silver Winners

12″ Compound Miter Saw • 777 Flight Deck • 7780 Workstation • Accumet pH Meter • Adjustable Keyboard • Casting Reel • CL2000 Disk Array Office Tower • Convertible Pen-Notebook Computer • Frankfurt Furniture Showroom • Future Generation ATM *(concept)* • GT Series Automotive Loudspeaker & Amplifier Packaging • IMPRAlert • IR2131 Impact Wrench • Leaf Eater • Lens Analyzer 350 • One-Piece Fishing Pliers • Open Office Enclosure Family • Orbit Lawnmower *(concept)* • Origins Retail Stores Program • Orion AW Photography Backpack • The Power of Maps Exhibition • Quest • Spectrum Analyzer • Spinal Cord Injury Patient Prone Cart *(concept)* • Trio Vacuum • Tristander

Bronze Winners

440 Personal Communicator • Activity Table • Anatomy Immersion Dissecting Table • Computer Accessory Product Range • Cordless HandyMixer • Cordless Mulching Mower • Cruiser/Commuter *(concept)* • DAL Fixtures • Digitizing Scanners • Display at Orgatec • DNA Sequencer • Door Positioning Hinge • E2 *(concept)* • Endoscope *(concept)* • Family CPR Trainer • Gig Saw *(concept)* • Insyte® Saf-T-Cath • Laserwriter Pro 600 & Pro 630 • Lightwave Integrated Network Cross-Connect System (LINXS) • MAX/it™ Desktop Terminal • Monet *(concept)* • Playback • Precision Toothbrush • Probe GT *(concept)* • Quantum • Scanjet IIc • SignTrack™ • Storage Express • System 3000 Models 3225/3230/3333 • TestBook • Ultrasonic Toothbrush • VCR VR5802/VR5602 • Versajet Water Jet Cutting System *(concept)* • Z-System

A + H International, Inc.
737 Bishop Street, Suite 2400
Honolulu, HI 96813

A + O Studio
1131 Tennessee Street
San Francisco, CA 94107

Tark Abed
c/o GVO Inc.
2370 Watson Court
Palo Alto, CA 94303

Able
23 West 35th Street
New York, NY 10001

ACS (Advanced Cardiovascular Systems)
3200 Lakeside Drive
Santa Clara, CA 95052

Acuson Corp.
1220 Charleston Road
Mountain View, CA 94043

Adam Straus Co.
321 West 29th Street, Third Floor
New York, NY 10001

AdStar, An IBM Company
9000 South Rita Road
Tucson, AZ 85744

Airtronics, Inc.
11 Autry
Irvine, CA 92718

Alan Stone Creative Services
16600 Sherman Way
Van Nuys, CA 91406

Althin Drake Willock
13520 Southeast Pheasant Court
Portland, OR 97222

American Honda Motors Co.
1919 Torrance Boulevard
Torrance, CA 90501

American Tourister, Inc
91 Main Street
Warren, RI 02885

Ampex Corporation
401 Broadway, MS 3-50
Redwood City, CA 94063

Anderson Design Associates
270 Farmington Avenue
Farmington, CT 06032

Anderson Warner
220A Division Street
Pleasanton, CA 94566

APCO
388 Grant Street Southeast
Atlanta, GA 30312

Apple Computer, Inc.
20730 Valley Green Drive, MS: 65-1D
Cupertino, CA 95014

Apple Computer, Inc.
10455 Bahdley Drive
Cupertino, CA 95014

Apple Computer, Inc.
20525 Mariani Avenue
Cupertino, CA 95014

Apple Computer, Inc.
3565 Monroe Street, 67-F
Santa Clara, CA 95051

Art Center College of Design
1700 Lida Street
Pasadena, CA 91103

Ashcraft Design
11832 West Pico Boulevard
Los Angeles, CA 90064

AT&T
5 Woodhollow Road
Parsippany, NJ 07054

Auburn University
Department of Industrial Design
O.D. Smith Hall, Room #103
Auburn, AL 36849

Audiovox Corp.
150 Marcus Boulevard
New York, NY 11788

Badsey Design of California
34102 Doheny Park Road
Capistrano Beach, CA 92624

Bally Design Inc.
424 North Craig Street
Pittsburgh, PA 15213

Baxter Healthcare Corp./Novacor Division
7799 Pardee Lane
Oakland, CA 94621

BBID
20800 Canon Drive
Los Gatos, CA 95030

Becton Dickinson Vascular Access
9450 South State Street
Sandy, UT 84070

Bell Northern Research
PO Box 3511, Station C
Ottawa, Ontario K1Y 4HY, Canada

Bennington Leather, Inc.
PO Box 787
North Bennington, VT 05257

Birtcher Medical Systems
50 Technology Drive
Irvine, CA 92718

Bissell Inc.
2345 Walker Northwest
Grand Rapids, MI 49504

Black & Decker (US) Inc.
701 East Joppa Road
Towson, MD 21286

Black & Decker Inc.
6 Armstrong Road
Shelton, CT 06484

Clint Blasingame
c/o J.L. Troupe Engineering & Design
S910A Commercial Drive
Huntsville, AL 35816

Bleck Design Group
139 Billerica Road
Chelmsford, MA 01824

Blount Corp.
4909 Southeast International Way
Portland, OR 97222

Boeder Design
1431 Chateau Common
Livermore, CA 94550

The Boeing Airplane Company
PO Box 3707, MS 02-JM
Seattle, WA 98124

Boston Whaler, Inc.
4121 South US Highway 1
Edgewater, FL 32141

Boston Whaler, Inc.
1149 Hingham Street
Rockland, MA 02370

Braund Creative Design, Inc.
601 Fourth Street, #120
San Francisco, CA 94111

Brayton International Coll.
255 Swathmore Avenue
High Point, NC 27264

Bresslergroup
1309 Noble Street
Philadelphia, PA 19123

Brion Vega
via Fratelli Gabba #9
Milan, Italy

Brown Jordan
9860 Gidley Street
El Monte, CA 91731

Browning-Ferris Industries, Inc.
150 Almaden Boulevard, Ninth Floor
San Jose, CA 95113

The Burdick Group
35 South Park
San Francisco, CA 94107

Burnes of Boston
582 Great Road
North Smithfield, RI 02895

Busse Hospital Disposables
PO Box 11067
Hauppauge, NY 11788

Cadillac Motor Division/General Motors Corp.
2860 Clark Street
Detroit, MI 48232

Caere Corp.
100 Cooper Court
Los Gatos, CA 95030

California State University–Long Beach
1250 Bellflower
Long Beach, CA 90815

California State University–Northridge
18111 Nordhoff
Northridge, CA 91330

Donald Carr
c/o NCR Corp.
1700 South Patterson Boulevard, PCD 4E
Dayton, OH 45479

Carter Shades
18475 Olympic Avenue South
Seattle, WA 98188

Center for Creative Studies
245 East Kirby
Detroit, MI 48202

Cary R. Chow
3699½ South Victoria
Los Angeles, CA 90016

Chrysler Corporation
800 Chrysler Drive East,
CIMS 483-56-02
Auburn Hills, MI 48326

Chrysler Corporation
12000 Chrysler Drive
Detroit, MI 48288

Coherent, Inc.
3270 West Bayshore Road
Palo Alto, CA 94303

Donna Cohn
4603 Schenley Road
Baltimore, MD 21210

ComputerCraft
1616 South Voss, Suite 100
Houston, TX 77057

CoStar Corp.
22 Bridge Street
Greenwich, CT 06830

Cranbrook Academy of Art
PO Box 801
Bloomfield Hills, MI 48303

Crawford Knives
205 North Center
West Memphis, AZ 72301

Crown Equipment Corp.
40-44 South Washington Street
New Breman, OH 45869

Dakota Jackson Inc.
306 East 61st Street
New York, NY 10021

Danaher Corp.
1250 24th Street Northwest, Suite 800
Washington, DC 20037

Data General Corp.
4400 Computer Drive
Westboro, MA 01580

De brey Design
6014 Blue Circle Drive
Minnetonka, MN 55343

Dell Computer
9505 Arboretum Boulevard
Austin, TX 78759

Deloitte & Touche
Two Oliver Plaza
Pittsburgh, PA 15222

Deloitte & Touche
28 Elmerest Drive
Wheeling, WV 26003

Deni/Keystone
20 Norris Street
Buffalo, NY 14207

Design Central
68 West Whittier
Columbus, OH 43206

The Design Company
501 Greenwich Street
San Francisco, CA 94133

Design Consortium
500 Park Overlook
Worthington, OH 43085

Design Continuum Inc.
648 Beacon Street
Boston, MA 02215

Design Edge
1202 San Antonio Street
Austin, TX 78701

Design for Industry
341 Linwood Avenue
Buffalo, NY 14209

Design Ideas
2500 Stockyard Road
Springfield, IL 62702

Design Solutions
73 East Elm Street
Chicago, IL 60611

Designhaus, Inc.
911 Western Avenue, Suite 555
Seattle, WA 98104

Designology Inc.
7641 East Gray Road
Scottsdale, AZ 85260

Designspring, Inc.
60 Post Road West
Westport, CT 06880

Designwerks!
1642 Las Trampas Road
Alamo, CA 94507

Designworks/USA
2201 Corporate Center Drive
Newbury Park, CA 91320

Digital Equipment Corp.
146 Main Street, MI011-3/L12
Maynard, MA 01754

Dusco Property Management
146 Park City Center
Lancaster, PA 17601

Emilio Ambasz Design Group
636 Broadway
New York, NY 10012

Empire Brushes, Inc.
PO Box 1606
Greenville, NC 27835

ERCO Leuchten GmbH
Postfach 2460
D-5880 Lündenscheid
Germany

Ergo Computing, Inc.
One Intercontinental Way
Peabody, MA 01960

Brian C. Ewing
2042 Stratford Avenue
South Pasadena, CA 91030

Exeter Architectural Products
243 West Eighth Street
Wyoming, PA 18644

Fahnstrom/McCoy Design
935 West Chestnut, Suite 305
Chicago, IL 60622

FAX Plus Inc.
2833 Bunker Hill Lane
Santa Clara, CA 95054

Fisher-Price, Inc.
636 Girard Avenue
East Aurora, NY 14052

Fiskars Inc.
7811 West Stewart Avenue
Wausau, WI 54401

Fitch, Inc.
(Formerly Fitch RichardsonSmith)
10350 Olentangy River Road
PO Box 360
Worthington, OH 43085

Fitness Master, Inc.
504 Industrial Boulevard
Waconia, MN 55387

Flanders + Associates
368 Congress Street
Boston, MA 02210

Ford Motor Company
21175 Oakwood Boulevard
Dearborn, MI 48123

Forest City Enterprises
10800 Brookpark Road
Cleveland, OH 44130

Formica Corp.
10155 Reading Road
Evendale, OH 45241

Formica Corp.
1501 Broadway, Suite 1519
New York, NY 10036

frogdesign inc.
4600 Bohannon Drive, Suite 101
Menlo Park, CA 94025

Fujitsu Limited
1015 Kamikodanaka, Nakahara-ku
Kawasaki 211, Japan

Frank J. Fusco
490 12th Street
Brooklyn, NY 11215

General Binding Corp.
One GBC Plaza
Northbrook, IL 60062

General Electric Plastics
One Plastic Avenue
Pittsfield, MA 01201

General Motors Corp.
3044 West Grand Boulevard
Detroit, MI 48202

General Motors Design Staff
30100 Mound Road,
GM Technical Center
Warren, MI 48090

General Motors/Hughes Aircraft
3050 West Lomita Boulevard
Torrance, CA 90505

George Podd Design
1440 North Dayton Street
Chicago, IL 60622

Gerry Baby Products Co.
12520 Grant Drive
Denver, CO 80241

The Gillette Company
1 Gillette Park, 2F-2
Boston, MA 02127

The Goodyear Tire and
Rubber Co.
Goodyear Technical Center,
PO Box 3531
Akron, OH 44309

The Goodyear Tire and
Rubber Co.
1144 East Market Street
Akron, OH 44316

GRiD Systems Corp.
47211 Lakeview Boulevard
Fremont, CA 94537

Group Four Design
147 Simsbury Road
Avon, CT 06001

GVO, Inc.
2470 Embarcadero Way
Palo Alto, CA 94303

Vincent L. Haley
2176 Good Road
Orrville, OH 44667

Handler
225 Varick, Ninth Floor
New York, NY 10014

Herbst Lazar Bell Inc.
345 North Canal Street
Chicago, IL 60606

Herman Miller, Inc.
8500 Byron Road
Zeeland, MI 49464

Hewlett-Packard
700 71st Avenue
Greeley, CO 80634

Hewlett-Packard
PO Box 301, MS 325
Loveland, CO 80537

Hewlett-Packard
1801 Page Mill Road, Building 18DD
Palo Alto, CA 94304

Hewlett-Packard
1266 Kifer Road, MS 102I
Sunnyvale, CA 94086

Hewlett-Packard Singapore Ltd.
1150 Depot Road
Singapore 0410

Todd Hoehn
1825 East Lafayette Place, #R
Milwaukee, WI 53202

Howe Furniture Corp.
12 Cambridge Drive
Trumbull, CT 06611

Huck and Studer Design, Inc.
17 Mendham Road
Gladstone, NJ 07934

Human Factors Industrial
Design, Inc.
575 Eighth Avenue
New York, NY 10018

Hycor Biomedical, Inc.
7272 Chapman Avenue
Garden Grove, CA 92641

IBM Corp.
11400 Burnet Road
Internal Zip 2950
Austin, TX 78758

IBM Corp.
1000 Northwest 51st Street, #3401
Boca Raton, FL 33429

IBM Corp.
740 New Circle Road
Lexington, KY 40511

ICI Americas, Inc.
Delaware Corporate Center
Wilmington, DE 19892

ID TWO
1527 Stockton Street
San Francisco, CA 94133

IDE Inc.
269 Mount Hermon Road
Scotts Valley, CA 95066

Ideations, design inc.
4257 24th Avenue West
Seattle, WA 98199

IDEO Product Development
660 High Street
Palo Alto, CA 94301

Igloo Products Corp.
1001 West Sam Houston
Parkway North
Houston, TX 77043

Ingersoll-Rand/Power Tools
Division
Allen & Martinsville Road
Liberty Corner, NJ 07938

INNO DESIGN, INC.
1290 Oakmead Parkway, Suite 310
Sunnyvale, CA 94086

Innovative Technologies
PO Box 90086
Houston, TX 77290

Interform
3475 Edison Way, Building F
Menlo Park, CA 94025

International Business
Machines Corp.
250 Harbor Drive, 2L-28
Stamford, CT 06904

Inventure Development Corp.
1308 Devils Reach Road, #302
Woodbridge, VA 22192

Inventure Development Corp.
Sales and Distribution Center
17800 South Main Street
Gardena, CA 90248

Island Design
153 Brace Road
West Hartford, CT 06107

The Jacobs Chuck
Manufacturing Co.
PO Box 592, One Jacobs Road
Clemson, SC 29633

James River Corp.
145 James Way
Southampton, PA 18966

Jarke-Thorsen Products, Inc.
925 South Capital Texas Highway,
Suite 245
Austin, TX 78746

JBL International
8500 Balboa Boulevard
Northridge, CA 91329

Jerome Caruso Design Inc.
37 Sherwood Terrace, Suite 123
Lake Bluff, IL 60044

JLF Designs, Inc.
PO Box 432
North Bennington, VT 05257

John Fluke Corp.
PO Box 9090
Everett, WA 98206

Joseph A. Wetzel Associates, Inc.
77 North Washington Street
Boston, MA 02114

Kaltenback & Voight
GmbH & Co.
KaVo Innovations
Bismarckring 39
D-7950 Biberach/Riss 1, Germany

Kemnitzer Design, Inc.
10880 Benson Drive, Suite 2300
Overland Park, KS 66210

Kennedy Design
28 South Park
San Francisco, CA 94107

Glenn Klaus
2 Brown Harrison
Jamaica Plain, MA 02130

Klitsner Indutrial Design Inc.
636 Fourth Street
San Francisco, CA 94107

The Knoll Group
655 Madison Avenue
New York, NY 10021

Kohler Co.
444 Highland Drive, MS 076
Kohler, WI 53044

Kozu Design
318 Lee Avenue
Yonkers, NY 10705

La Croix Water Co.
100 Harborview Plaza
La Crosse, WI 54601

Laerdal California, Inc.
1901 Obispo Avenue
Long Beach, CA 90804

Landor Associates
230 Park Avenue South, Seventh Floor
New York, NY 10003

Leap Inc.
111 East 16th Avenue
Eugene, OR 97401

Lee Company
9001 West 67th Street
Merriam, KS 66202

Levolor Corp.
595 Lawrence Expressway
Sunnyvale, CA 94086

Logitech Inc.
6505 Kaiser Drive
Fremont, CA 94555

Lou Hammond & Associates
39 East 51st Street, Fourth Floor
New York, NY 10022

Lowepro/Fluxion
2194 Northpoint Parkway
Santa Rosa, CA 95407

Lumatech
5515 Doyle Street
Emeryville, CA 94608

Lunar Design
119 University Avenue
Palo Alto, CA 94310

Lunar Design
537 Hamilton Avenue
Palo Alto, CA 94301

Machen Montague
2221 Edge Lake Drive, Suite 100
Charlotte, NC 28217

Magna Design Corp.
1740 Technology Drive
San Jose, CA 95110

Marc Harrison Associates
568 Bristol Ferry Road
Portsmouth, RI 02871

Martin Marietta Energy
Systems Inc.
PO Box 2003, Building K-1225
Oak Ridge, TN 37831

Matrix Product Design, Inc.
660 High Street
Palo Alto, CA 94301

Matrix Tool & Mold Co.
358 Warehouse Road
Oak Ridge, TN 37830

McBrayer Industrial Design
PO Box 310186
New Braunfels, TX 78131

Medtronic, Inc.
7000 Central Avenue
Minneapolis, MN 55432

Metaphase Design Group
1266 Andes Boulevard
St. Louis, MO 63132

The Metraflex Company
2323 West Hubbard Street
Chicago, IL 60612

Microsoft Corp.
One Microsoft Way
Redmond, WA 98052

Millipore Corp.
80 Ashby Road
Bedford, MA 01730

Momenta Corp., Inc.
295 North Bernardo Avenue
Mountain View, CA 94043

Moss, Inc.
Mount Battie Street
Camden, ME 04843

MOTOcycles
11625 Norwood Road
Raleigh, NC 27613

Museum of Contemporary Art,
San Diego
1001 Kettner Boulevard
San Diego, CA 92101

Museum of Science and Industry
57th and Lake Shore Drive
Chicago, IL 60637

Navistar International
455 North Cityfront Plaza
Chicago, IL 60611

Navistar International
2911 Meyer Road
Fort Wayne, IN 46803

NCR Corp.
1700 South Patterson
Boulevard, PCD4E
Dayton, OH 45479

nCube
1825 Northwest 167th Place
Beavertown, OR 97006

Nissan Design International, Inc.
9800 Campus Point Drive
San Diego, CA 92121

North Carolina State University
4700 Hillsborough Street
Raleigh, NC 27695

Northern Telecom
8614 Westwood Court Drive
Vienna, VA 22182

Northwest Airlines
5101 Northwest Drive
Department A6800
St. Paul, MN 55111

Novax Handcrafted Guitars
638 Dowling Boulevard
San Leandro, CA 94577

O'Donnell Pet Products
2501 Pepperwood
Long Beach, CA 90815

O'Neill, Inc.
1071 41st Avenue, Box 6300
Santa Cruz, CA 95063

Ohio State University
1790 North Fort Street
Columbus, OH 43201

Oxo International
15 Gramercy Park South
New York, NY 10003

Palo Alto Design Group
360 University Avenue
Palo Alto, CA 94301

Patton Design
8 Pasteur #170-1
Irvine, CA 92718

Pelican Design
112A South Danby Road
Spencer, NY 14883

Pentagram Design
212 Fifth Avenue
New York, NY 10010

PerSeptive BioSystems
38 Sidney Street
Cambridge, MA 02139

Peter Forbes and Associates, Inc.
241 A Street
Boston, MA 02210

Pinnacle Systems
2380 Walsh
Santa Clara, CA 95051

PlayDesigns
PO Box 505
315 Cherry Street
New Berlin, PA 17855

Poqet Computer Corp.
5200 Patrick Henry Drive
Santa Clara, CA 95054

Pratt Institute
200 Willoughby Avenue
Brooklyn, NY 11205

Precision Mfg. Inc.
2200 52nd Avenue
Lachine, Quebec H8T 2Y6, Canada

Preschool R & D, Playskool
Division Of Hasbro
1027 Newport Avenue
Pawtucket, RI 02861

Product Insight, Inc.
6 Ledgerock Way, Unit 1
Acton, MA 01720

Product Solutions Inc.
447 New Grove Street
Wilkes-Barre, PA 18702

Proformix Inc.
Route 22 West
Whitehouse Station, NJ 08883

Purdue University
Industrial Design Department
West Lafayette, IN 47907

QNET
6100 Executive Boulevard, Suite 300
Rockville, MD 20852

Qualtronics
433 Caredean Drive
Horsham, PA 19044

Rauscher Design Inc.
211 McCready Avenue
Louisville, KY 40206

Reebok USA
100 Technology Drive
Stoughton, MA 02072

Tim C. Repp
7 Juniper Circle
Canton, CT 06019

Ricchio Design
PO Box 3028
Seal Beach, CA 90740

Rinehart Glove, Ltd.
PO Box 12377
Aspen, CO 81612

Brett D. Ritter
925 Coastline Drive
Seal Beach, CA 90740

Roche Harkins Inc.
17 Clinton Drive
Hollis, NH 03049

Rollerblade Inc.
5101 Shady Oak Road
Minnetonka, MN 55343

RTKL Associates
2828 Routh Street, #200
Dallas, TX 75201

Rycom Instruments, Inc.
951 East 59th
Raytown, MO 64133

Ryobi Motor Products Corp.
1424 Pearman Dairy Road
Anderson, SC 29625

Ryobi Motor Products Corp.
225 Pumpkintown Road, PO Box 35
Pickens, SC 29671

Saf-T-Loc
9904 Clayton Road
St. Louis, MO 63124

Jane Saks-Cohan
7, The Cloisters 11 Salem Road
London, W2 4DI, England

Samsonite
11200 East 45th Avenue
Denver, CO 80239

Schott/Surgitek
3037 Mount Pleasant Street
Racine, WI 53404

Scott/Sani-Fresh
International, Inc.
4702 Goldfield Drive
San Antonio, TX 78218

SDA
Hoeksteen 41-43
2132 Mount Hoofddorp,
The Netherlands

Sears Roebuck & Co.
Sears Tower
Chicago, IL 60684

Serbinski Machineart
66 Willow Avenue
Hoboken, NJ 07030

S.G. Hauser Associates
24009 Ventura Boulevard, Suite 200
Calabasas, CA 91302

Sherwood Medical Company
444 McDonnell Boulevard
Hazelwood, MO 63044

Siler/Siler Ventures
3328 Lakeview Boulevard
Lake Oswego, OR 97035

Skidmore, Owings & Merrill
333 Busk Street
San Francisco, CA 94104

Slam Design
634 A Venice Boulevard
Venice, CA 90291

Steven M. Slaton
16 Enterprise Street, #2
Raleigh, NC 27607

Smart Design Inc.
7 West 18th Street
New York, NY 10011

Martin Smith
c/o Art Center College of Design
1700 Lida Street
Pasadena, CA 91104

Soma, Inc.
514 Northwest 11th Avenue, #209
Portland, OR 97209

Sonneman Design Group, Inc.
26-11 Jackson Avenue
Long Island City, NY 11101

Sony Corporation of America
1 Sony Drive
Park Ridge, NJ 07656

Southco, Inc.
210 North Brinton Lake Road
Concordville, PA 19331

Space, Form & Structure
11625 Norwood Road
Raleigh, NC 27613

Spalding Sports Worldwide
425 Meadow Street
Chicopee, MA 01021

Spectralogic, Inc.
430 Tenth Street, S/206
Atlanta, GA 30318

Spectraphysics
959 Terry Street
Eugene, OR 97402

Sports-Mitt International
46 Shattuck Square, Suite 14
Berkeley, CA 94704

Square D Company
1601 Mercer Road
Lexington, KY 40511

Squejé
PO Box 328
Menlo Park, CA 94026

Staveley Sensors, Inc.
91 Prestige Park Circle
East Hartford, CT 06108

Steelcase Inc.
901 44th Street Southeast
Grand Rapids, MI 49508

Peter Stathis
15 Academy Street
Cold Spring, NY 10516

Steiner Design
214 Pemberwick Road
Greenwich, CT 06831

STUDIOS Architecture
1133 Connecticut Avenue Northwest,
Suite 500
Washington, DC 20036

Greg Stuhl
2207 North Ponderosa
Santa Ana, CA 92701

Sunstar, Inc.
3-1 Asahi-Machi, Takasuki
Osaka, Japan

Support Systems International,
Inc. (S.S.I.)
4349 Corporate Road
Charleston, SC 29405

Swany & US Army
417 Fifth Avenue, Room #1115
New York, NY 10016

Barry Sween
21462 PCH #288
Huntington Beach, CA 92648

Syva Company
900 Arastradero Road
Palo Alto, CA 94303

Tandem Computers, Inc.
10300 North Tantan Avenue, #55-54
Cupertino, CA 95014

Taylor & Chu
1831 Powell Street
San Francisco, CA 94133

TCB Inc., Concepts Plus Division
1227 East Hennelin Avenue
Minneapolis, MN 55414

Teague
14727 Northeast 87th Street
Redmond, WA 98052

TECHNO S.p.A.
via U1 6L1 22
Milan, Italy

The Technology Center of
Silicon Valley
145 West San Carlos Street
San Jose, CA 95113

Telect, Inc
PO Box 665
Liberty Lake, WA 99019

Terumo Corp.
2001 Cottontail Lane
Somerset, NJ 08873

Thermedics, Inc.
470 Wildwood Street
Woburn, MA 01888

Thinking Machines Corp.
245 First Street
Cambridge, MA 02142

Thomson Consumer Electronics
600 North Sherman Drive
Indianapolis, IN 46201

Thornet Industries
2001 Speedball Road
Statesville, NC 28677

Timesavers, Inc.
5270 Hanson Court
Minneapolis, MN 55429

Toledo Scale
350 West Wilson Bridge Road
Worthington, OH 43085

Trigem Corp.
2388 Walsh Avenue, Building B
Santa Clara, CA 95051

Tucker Housewares
25 Tucker Drive
Leominster, MA 01453

Tumble Forms Inc.
Barker Road
Dolgeville, NY 13329

u r o designs
7129 Emily Street, Northeast
Albuquerque, NM 87109

Ultradata
5020 Franklin Drive
Pleasanton, CA 94588

United Blood Services
6220 East Oak Street
Scottsdale, AZ 85257

University of Bridgeport
Bruell Hall, 600 University Avenue
Bridgeport, CT 06601

US Stamp
2150 Trabajo Drive
Oxnard, CA 93031

Vent Design
1436 White Oaks Road, #15
Campbell, CA 95008

Verifone Inc.
3 Lagoon Drive, Suite 400
Redwood City, CA 94065

Versteel
PO Box 850
Jasper, IN 47547

Vivitar Corporation
9350 DeSoto Avenue
Chatsworth, CA 91311

Walter Dorwin Teague Associates
14727 Northeast 87th Street
Redmond, WA 98052

Waters Design Associates Inc.
3 West 18th Street
New York, NY 10011

Wedge Innovations
2040 Fortune Drive
San Jose, CA 95131

West Office Design
Associates, Inc.
118 Hawthorne Street
San Francisco, CA 94107

Eric Williams
652 Hudson Street, #3E
New York, NY 10010

Winter Design Mfg.
1200 Pepperwood Drive
Danville, CA 94506

Wittenbrock Design
4815 Southwest Patton Road
Portland, OR 97221

Mary Heather Worley
5707 Bullard Drive
Austin, TX 78757

Xerox Corporation
1350 Jefferson Road
Rochester, NY 14623

Yamaha Corp.
10-1 Nakazawa
Hamamatsu, Japan

Yonca Teknik Yatirim A.S.
Tersaneler Caddessi, Gecici 50 Sokak
81700 Tuzla, Istanbul, Turkey

Zaca Investments, Inc.
3601 Forest Gate Circle
Malibu, CA 90265

Zelco Industries, Inc.
630 South Columbus Avenue
Mount Vernon, NY 10550

ZIBA Design, Inc.
305 Northwest 21st Avenue
Portland, OR 97210

INDEX

PRODUCTS

DESIGNERS

DESIGN FIRMS

DESIGN SCHOOLS

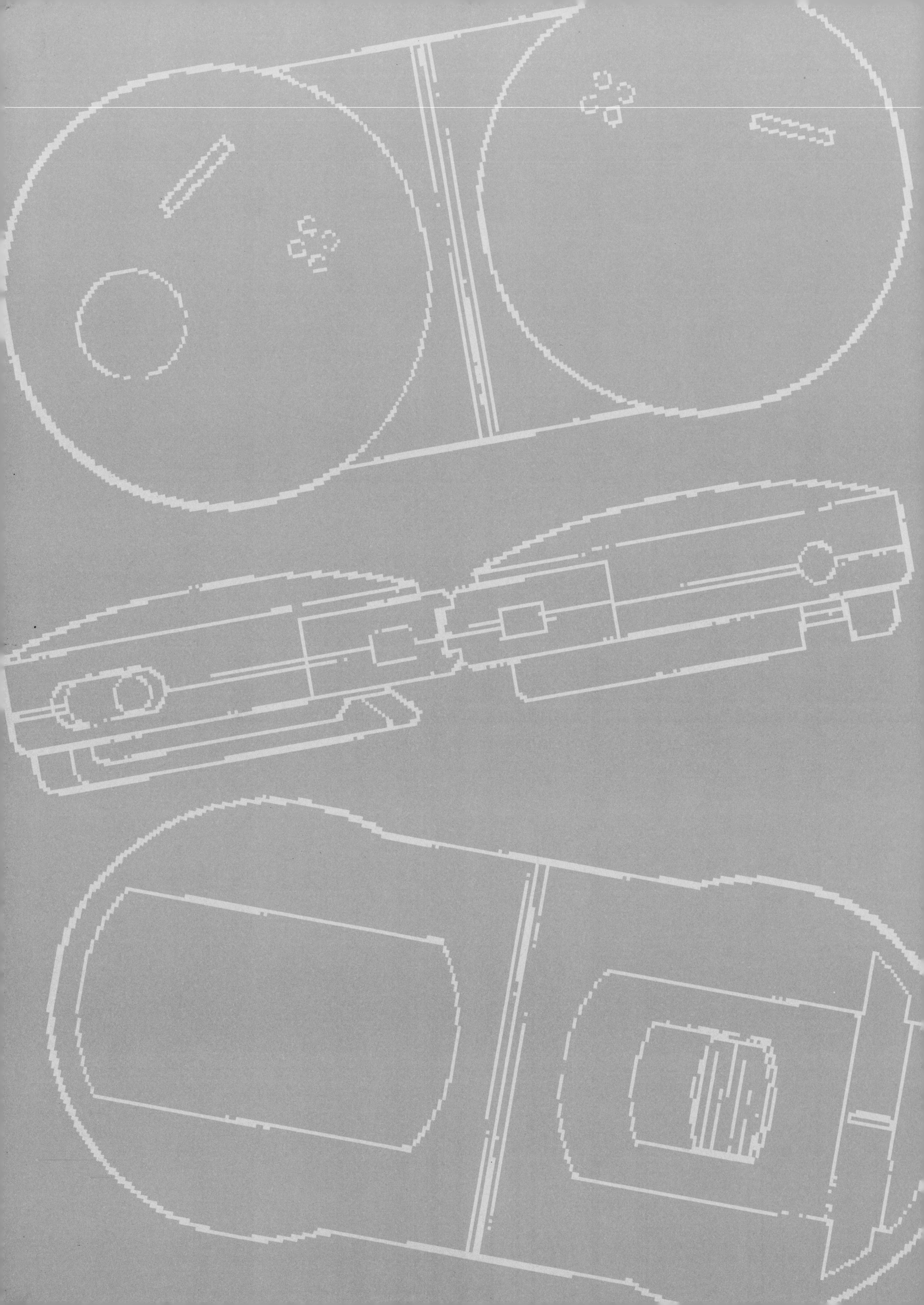